SpringerBriefs in Computer Science

Series Editors

Stan Zdonik
Shashi Shekhar
Jonathan Katz
Xindong Wu
Lakhmi C. Jain
David Padua
Xuemin (Sherman) Shen
Borko Furht
V. S. Subrahmanian
Martial Hebert
Katsushi Ikeuchi
Bruno Siciliano
Sushil Jajodia
Newton Lee

More information about this series at http://www.springer.com/series/10028

Hassan Aboubakr Omar • Weihua Zhuang

Time Division Multiple Access For Vehicular Communications

 Springer

Hassan Aboubakr Omar
University of Waterloo
Waterloo
Ontario
Canada

Weihua Zhuang
University of Waterloo
Waterloo
Ontario
Canada

ISSN 2191-5768   ISSN 2191-5776 (electronic)
ISBN 978-3-319-09503-5 ISBN 978-3-319-09504-2 (eBook)
DOI 10.1007/978-3-319-09504-2
Springer Cham Heidelberg New York Dordrecht London

Library of Congress Control Number: 2014946651

Printed on acid-free paper

Springer is part of Springer Science+Business Media (www.springer.com)

To my mother, Samia, my father, Aboubakr,
and my sister, Dina—H.A.O.

To my sons, Alan and Alvin—W.Z.

Preface

Road accidents represent a serious social problem and are one of the leading causes of human death and disability on a global scale. To reduce the risk and severity of a road accident, a variety of new safety applications can be realized through wireless communications among vehicles moving near each other, or among vehicles and especially deployed road side units (RSUs), a technology known as a vehicular ad hoc network (VANET). Most of the VANET-enabled safety applications are based on broadcasting of safety messages by vehicles or RSUs, either periodically or event-driven, such as in case of a hard brake or dangerous road condition detection. Each broadcast safety message should be successfully delivered to the surrounding vehicles and RSUs without any excess delay, which is one of the main functions of a medium access control (MAC) protocol in VANETs.

This brief presents VeMAC, a new multichannel time division multiple access (TDMA) protocol specifically designed to support the high priority safety applications in a VANET scenario. The ability of the VeMAC protocol to deliver periodic and event-driven safety messages in VANETs is demonstrated by a detailed delivery delay analysis, including queueing and service delays, for both types of safety messages. As well, computer simulations are conducted by using MATLAB, the network simulator ns-2, and the microscopic vehicle traffic simulator VISSIM, in order to evaluate the performance of the VeMAC protocol, in comparison with the IEEE 802.11p standard and the ADHOC MAC protocol (another TDMA protocol proposed for ad hoc networks). A real city scenario is simulated and different performance metrics are evaluated, including the network goodput, protocol overhead, channel utilization, service fairness, probability of a transmission collision, and safety message delivery delay. It is shown that the VeMAC protocol considerably outperforms the existing MAC schemes in delivering periodic and event-driven safety messages in VANETs.

The proposed VeMAC protocol can be applied for many advanced safety applications to enhance the public safety standards and improve the safety level of drivers/passengers and pedestrians on roads. This research sheds light on TDMA as a promising technology for MAC in VANETs, and a suitable replacement of the IEEE 802.11p standard, which has significant limitations in supporting VANET safety applications.

Waterloo, Ontario, Canada   Hassan Aboubakr Omar
June 2014   Weihua Zhuang

Acknowledgements

We would like to thank all our colleagues in the Broadband Communications Research (BBCR) Lab at the University of Waterloo for the beneficial discussions, research collaboration, and continuous exchange of knowledge. Also, we gratefully acknowledge Usama Shahdah, a PhD student in the Transportation Systems Research group at the University of Waterloo, for his very useful guidance for conducting vehicle traffic computer simulations.

Contents

List of Abbreviations

AcS   Acceptance of Services
AnS   Announcement of Services
HPSA   High Priority Short Applications
1PPS   One Pulse Per Second
A-opt   ADHOC-optimal
AC   Access Category
AE   ADHOC-enhanced
AIFS   Arbitrary Interfram Space
BSM   Basic Safety Message
CCH   Control Channel
CDF   Cumulative Distribution Function
CDMA   Code Division Multiple Access
CPThresh   Capture Threshold
CSThresh   Carrier Sensing Threshold
CTS   Clear-To-Send
CW   Contention Window
DSRC   Dedicated Short Range Communications
EDCA   Enhanced Distributed Channel Access
ETSI   European Telecommunications Standards Institute
FCC   Federal Communication Comission
FCS   Frame Check Sequence
GPS   Global Positioning System
HOL   Head of Line
i.i.d.   Independent and Identically Distributed
ID   Identifier
ITS   Intelligent Transportation System
LTE-A   Long Term Evolution Advanced
MAC   Medium Access Control
MPDU   MAC Layer Protocol Data Unit
MSDU   MAC Layer Service Data Unit
MTU   Maximum Transmission Unit
OBU   On-Board Unit

OFDM	Orthogonal Frequency Division Multiplexing
PGF	Probability Generating Function
PLCP	Physical Layer Convergence Procedure
PMF	Probability Mass Funciton
PN	Pseudo Noise
QoS	Quality-of-Service
RSU	Road Side Unit
RTS	Request-To-Send
RxThresh	Receiving Threshold
SCH	Service Channel
SDMA	Space Division Multiple Access
SRP	Slot Release Prevention
TDMA	Time Division Multiple Access
THS	Two-Hop Set
THSO	Two-Hop Set Occupancy
USDoT	United States Department of Transportation
UW	University of Waterloo
V-inf	VeMAC with $\tau = \infty$
V0	VeMAC with $\tau = 0$
V2R	Vehicle-To-Road Side Unit
V2V	Vehicle-To-Vehicle
VANET	Vehicular Ad Hoc Network
VSC	Vehicle Safety Communications

Nomenclature

λ	Arrival rate of event-driven safety messages
λ_v	Rate of vehicle arrival from each possible entry of the road network in the VISSIM simulations
$\mathcal{A}_x$	Set of time slots that node x can attempt to access on the control channel
$\mathcal{D}_x$	Set of one-hop neighbors of node x which have not successfully received a certain packet broadcasted on the control channel (possibly by node x itself)
$\mathcal{F}$	Set of time slots on the control channel associated with road side units
L	Set of time slots on the control channel associated with vehicles moving to the left directions
$\mathcal{N}_x$	Set of one-hop neighbours of node x, from which node x has received VeMAC Type1 packets on the control channel in the previous L slots
R	Set of time slots on the control channel associated with vehicles moving to the right directions
$\mathcal{T}_x$	Set of time slots that node x must not use on the control channel in the next L time slots
$\mathcal{T}_x^m$	Set of time slots that node x must not use on channel cm in the next lm time slots
E	Event (probability theory)
μ_n	Average number of contending nodes which acquire a time slot within n frames obtained from mathematical analysis
μ_n^{sim}	Average number of contending nodes which acquire a time slot within n frames obtained from computer simulations
$\bar{X}$	First moment of a random variable X
$\overline{X^2}$	Second moment of a random variable X
τ	VeMAC split up parameter
BX_{add}	Additive part of the safety distance
BX_{mult}	Multiplicative part of the safety distance
ID_k^x	VeMAC ID used by node x to access time slot k

Length_h	Length of the highway segment simulated in MATLAB
Length_s	Length of a street in the city scenario simulated in MATLAB
ϑ_x^m	Set of time slots over which a provider node x offers a service on service channel c_m
A_r^n	Number of ways of obtaining an ordered subset of r elements from a set of n elements
a_y^m	Time slot over which a user node x transmits an acknowledgement packet on service channel c_m
c_i	Service channel number i
C_r^n	Number of ways of obtaining a subset of r elements from a set of n elements
F_n	Probability that a specific node (among the contending ones) acquires a time slot within n frames
F_n^{all}	Probability that all the contending nodes acquire a time slot within n frames
F_X	Cumulative distribution function of a discrete random variable
f_X	Probability mass function of a discrete random variable
$G_X'(z)$	Derivative of $G_X(z)$ with respect to z
G_X	Probability generating function of a discrete random variable which takes non-negative integer values
I	Inter-arrival time of periodic safety messages
J	Index of a time slot at the start of which a safety message becomes the head of line
K	Total number of nodes which need to acquire a time slot on the control channel
k	Number of time slots that a node acquires per frame on the control channel
k_e	Number of event-driven time slots that a node acquires per frame on the control channel
k_p	Number of periodic time slots that a node acquires per frame on the control channel
L	Number of time slots per frame on the control channel
l_m	Number of time slots per frame on channel c_m
$M_{i,j}$	Entry of a matrix M, located at the ith row and j^{th} column
N	Number of available time slots in a frame on the control channel
N_c	Number of contending nodes attempting to acquire a specific time slot
N_f	Number of periodic safety message fragments queued before a tagged fragment within a batch
n_f	Total number of fragments of a periodic safety message
N_q	Total number of vehicles successfully acquiring a time slot
N_s	Total number of streets in the city scenario simulated in MATLAB
N_v	Total number of vehicles used for MATLAB simulations in highway and city scenarios

N_{sch}	Number of service channels
N_{max}	Maximum number of nodes which can exist in a two-hop set
N_{succ}^{x}	Number of nodes in the two-hop neighbourhood of node x which have successfully acquired a time slot
P	One-step transition probability matrix of a Markov chain
P^n	n-step transition probability matrix of a Markov chain
$p(X = x \mid Y = y)$	Conditional probability that discrete random variable X takes the value x given that discrete random variable Y takes the value y
p_{acc}	Probability of accessing an available time slot in the ADHOC MAC protocol
p_{ij}	Transition probability from state i to state j of a Markov chain
p_{opt}	Optimal probability of accessing an available time slot
R	Communication range
r_x	Ratio of the number of safety messages transmitted by node x to the total number of safety messages transmitted by all nodes during the simulation time
s_x	Total number of safety messages generated at node x normalized by the total number of safety messages generated at all nodes during the simulation time
t	Duration of a time slot on the control channel
t_{in}	Total time of arrival of vehicles to the road network in the VISSIM simulations
t_w	Warm up period in the VISSIM simulations
W	Total delay (sum of the service delay and queueing delay)
$W_{(l, u, v)}$	Number of ways by which l nodes can acquire a time slot given that there are u contending nodes each randomly choosing a time slot among v available time slots
W_b	Total service delay of a batch of fragments of a periodic safety message
W_q	Queueing delay
W_s	Service delay
W_{q1}	Delay since a batch of fragments of a periodic safety enters the queue until the first fragment of the batch becomes the head of line
W_{q2}	Service delay of all the periodic safety message fragments queued before a tagged fragment within a batch
X_n	Total number of nodes which acquired a time slot within n frame
AX	Average standstill distance
$\lvert \mathcal{X} \rvert$	Cardinality of a set $\mathcal{X}$

Chapter 1
Introduction

1.1 Vehicular Ad Hoc Networks

An ad hoc network is defined as a collection of nodes dynamically forming a network without any existing infrastructure or centralized administration. One special type of mobile ad hoc networks is the network among moving vehicles, which is known as vehicular ad hoc network (VANET). A VANET is an emerging technology which consists of a set of vehicles, each equipped with a communication device called on-board unit (OBU), and a set of stationary units along the road, referred to as road side units (RSUs). As shown in Fig. 1.1, some RSUs can act as a gateway for connectivity to other communication networks, such as the Internet. Each vehicle OBU has a wireless network interface which allows the vehicle to directly connect to other vehicles and RSUs within its communication range, as well as a wireless or wired interfaces to which application units can be attached. By employing vehicle-to-vehicle (V2V) and vehicle-to-RSU (V2R) communications, VANETs can support a wide variety of applications in road safety, passenger infotainment, and vehicle traffic optimization [4–6], which is the main reason that VANETs have received significant support from government, academia, and industrial organizations over the globe.

Motivated by the importance of vehicular communications, in 1999, the United States Federal Communication Comission (FCC) has allocated 75 MHz radio spectrum in the 5.9 GHz band for dedicated short range communications (DSRC) to be exclusively used by V2V and V2R communications. Similarly, in 2008, the European Telecommunications Standards Institute (ETSI) has allocated 30 MHz of spectrum (also in the 5.9 GHz band) for Intelligent Transportation System (ITS) applications. As shown in Fig. 1.2, the DSRC spectrum is divided into seven 10 MHz channels: six service channels for safety and non-safety related applications, and one control channel for transmission of control information and high priority safety messages. Such allocation of radio spectrum for vehicular communications has motivated the establishment of many national and international research projects, e.g., [8–17],

© The Author(s) 2014   1
H. A. Omar, W. Zhuang, *Time Division Multiple Access For Vehicular Communications*,
SpringerBriefs in Computer Science, DOI 10.1007/978-3-319-09504-2_1

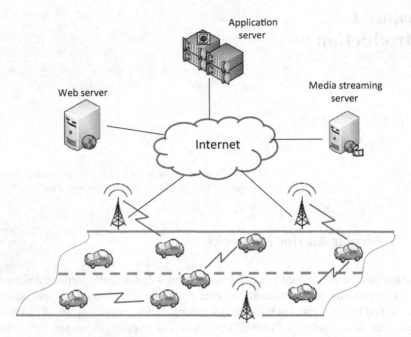

Fig. 1.1 Illustration of a VANET

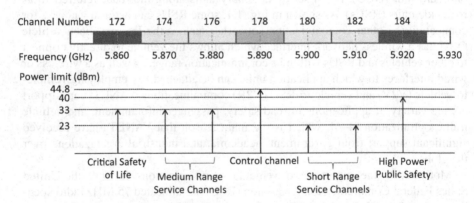

Fig. 1.2 The DSRC spectrum allocated by the FCC, with the effective isotropically radiated power (EIRP) limits as specified in the ASTM E2213 standard [7] for public RSUs

which are all dedicated to the research in VANETs. Every project has its unique objectives, focusing on safety related applications [14, 15], security of vehicular communications [10, 11], the development of a simulation platform for V2V and V2R communications [17], and so on.

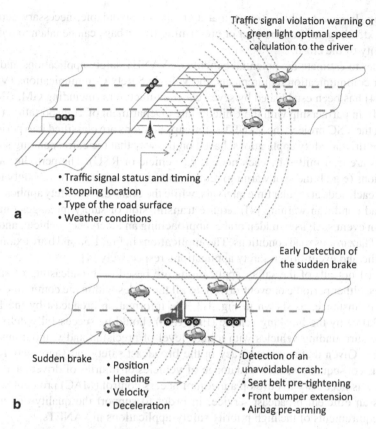

Fig. 1.3 Examples of safety applications based on V2R (Fig. 1.3a) and V2V (Fig. 1.3b) communications as defined in [4]. **a** Traffic signal violation warning and green light optimal speed advisory. **b** Emergency electronic brake light and pre-crash sensing

1.2   VANET Safety Applications

The primary category of VANET applications is to enhance the public safety standards and provide a safer environment for drivers/passengers and pedestrians on road. For instance, at a signalized intersection as shown in Fig. 1.3a, an RSU can continuously broadcast to the approaching vehicles information about the traffic signal status and timing, stopping location, type of road surface, weather conditions, etc. Then, based on this broadcast information, the in-vehicle system can predict a traffic signal violation and give a warning to the driver, or advise him/her with an optimal speed to reach the traffic signal during the green light phase. Figure 1.3b illustrates examples of safety applications that are based on V2V communications. As shown in Fig. 1.3b, if a vehicle suddenly breaks, it broadcasts information about its current status (i.e., position, speed, deceleration, etc.), which is used by the surrounding vehicles to early detect the sudden brake, even in limited visibility conditions, such as due to heavy fog. In case a vehicle senses that a crash is unavoidable, necessary actions such as extending the front bumper or pre-arming the airbags can be taken to reduce the severity of the crash.

In order to estimate the potential benefits of VANET safety applications and define their communication requirements, the Vehicle Safety Communications (VSC) project [4] has been established by seven car manufacturers (including GM, BMW, and Ford), in partnership with the United States Department of Transportation (US-DoT). In the VSC project, the VANET safety applications are classified into periodic and event-driven safety applications, based on the way that the corresponding safety messages are transmitted by each node (i.e., vehicle or RSU). The periodic safety applications (e.g., blind spot warning) require automatic transmission of safety messages by each node at regular time intervals, while the event-driven safety applications (e.g., road condition warning [4]) require transmission of safety messages only in case of an event such as a sudden brake, approaching an emergency vehicle, and detection of hazardous road conditions. The applications in Fig. 1.3a and b are examples of periodic and event-driven safety applications respectively [4].

Most (if not all) of the safety applications are based on broadcasting of safety messages, either periodic or event-driven, to all the nodes within the communication range. For instance, as shown in Fig. 1.3, the information broadcast by the RSU (Fig. 1.3a) or by the breaking vehicle (Fig. 1.3b) should be successfully delivered to all the surrounding vehicles with a high level of precision and without any excess delay. Given that any inaccuracy in the broadcast safety messages may result in serious consequences, such as damage of vehicles or injuries of drivers and passengers, it is necessary to develop a medium access control (MAC) protocol which provides an efficient broadcast service, in order to support the quality-of-service (QoS) requirements of the high priority safety applications in VANETs.

1.3   MAC in VANETs

Various MAC protocols have been proposed for VANETs, based either on IEEE 802.11 [18, 19] or on channelization such as space division multiple access (SDMA), code division multiple access (CDMA), and time division multiple access (TDMA). In SDMA schemes, each vehicle decides whether or not it is allowed to access the channel based on its location on the road [20, 21]. An SDMA scheme consists of the following three components. First, a *discretization procedure* divides the road into small areas called cells. Each cell may contain one [20] or more vehicles [21] based on the size of the cell determined by the SDMA scheme. Second, a *mapping function* assigns to each of the cells a unique time slot. To avoid the hidden terminal problem, two cells are assigned the same time slot only if the distance between them is greater than twice of the communication range. Third, an *assignment rule* specifies which time slots a vehicle is allowed to access based on its current location. For any SDMA scheme, the vehicles should be able to correctly determine their current position, store the mapping of the cells into time slots, and synchronize to a common reference. One main problem of SDMA schemes is that, when most of the cells on a road are unoccupied by vehicles, the unused time slots assigned to these cells represent a waste of bandwidth.

Similarly, CDMA is proposed for MAC in VANETs due to its robustness against interference and noise [22, 23]. The main issue which arises with CDMA in VANETs is how to allocate the pseudo noise (PN) codes to different vehicles. Due to a large number of vehicles, if every vehicle is assigned a unique PN code, the length of these codes will become extremely long, and the required bit rates for VANET applications may not be attained. Consequently, it is mandatory that the PN codes be shared among different vehicles in a dynamic and fully distributed way. One solution for distributed PN code allocation is to provide each node with several filters matched to the available PN codes, which are shared by all nodes [22]. Then, each node attempts to select a PN code that is not used by other nodes within its communication range, by selecting the PN code corresponding to the matched filter which gives the minimal output. However, this scheme suffers from the hidden terminal problem, and its implementation is impractical due to its requirement of a matched filter for each PN code. Although the number of matched filters can be reduced by using a location based PN allocation scheme [23], the complexity of these matched filters increases since they need to be adaptive. That is, each filter should match to different PN codes based on the area where the vehicle is currently located. The complexity of implementation is the main disadvantage of CDMA schemes.

The ADHOC MAC is the most well known TDMA protocol proposed for inter-vehicle communication networks [24–29]. The ADHOC MAC protocol operates in a time slotted structure, where time slots are grouped into virtual frames, i.e., no frame alignment is needed. By letting each node report the status of all the time slots in the previous (sliding) virtual frame, the ADHOC MAC can support an acknowledged broadcast service without the hidden terminal problem [24]. However, the ADHOC MAC protocol has major limitations which significantly degrades its performance, as will be discussed in details in Sect. 3.5.3. On the other hand, the main solution currently proposed for MAC in VANETs is the IEEE 802.11 p standard [30]. The protocol is based on the legacy IEEE 802.11 standard, which is widely implemented, with new parameter values for the enhanced distributed channel access (EDCA) [30, 18] scheme to be used for communications over the control channel (CCH) (as recommended by the IEEE 1609.4 standard [31]). However, as will be explained in Sect. 4.3, the IEEE 802.11 p standard does not provide a reliable broadcast service, which considerably reduces its ability to support the periodic and event-driven safety applications in VANETs.

1.4 Brief Objective and Outline

Motivated by the limitations of the MAC schemes proposed for VANETs, as discussed in Sect. 1.3, this brief introduces a multichannel TDMA MAC protocol which can provide a reliable one-hope broadcast service, necessary to support the QoS requirements of VANET periodic and event-driven safety applications. The rest of the brief is organized as follows. Chapter 2 describes the system model under consideration, while Chap. 3 introduces a novel multichannel TDMA protocol, called VeMAC, and compares its performance with that of ADHOC MAC, via simulations in highway and city scenarios [1, 32, 33]. The ability of the VeMAC protocol to support periodic and event-driven safety messages is investigated in Chap. 4 in comparison with the IEEE 802.11 p standard [2, 3]. This performance evaluation is done by presenting a detailed delivery delay analysis for periodic and event-driven safety messages, and conducting computer simulations in a realistic city scenario consisting of roads around the University of Waterloo (UW). Finally, Chap. 5 concludes this brief and suggests some further research topics.

References

1. H. A. Omar, W. Zhuang, and L. Li, "VeMAC: A TDMA-based MAC protocol for reliable broadcast in VANETs," *IEEE Transactions on Mobile Computing*, vol. 12, no. 9, pp. 1724–1736, Sept. 2013.
2. H. A. Omar, W. Zhuang, A. Abdrabou, and L. Li, "Performance evaluation of VeMAC supporting safety applications in vehicular networks," *IEEE Transactions on Emerging Topics in Computing*, vol. 1, no. 1, pp. 69–83, Jun. 2013.
3. H. A. Omar, W. Zhuang, and L. Li, "Delay analysis of VeMAC supporting periodic and event-driven safety messages in VANETs," in *Proc. IEEE Global Communications Conference (GLOBECOM)*, Dec. 2013.
4. "Vehicle safety communications project task 3 final report," The CAMP Vehicle Safety Communications Consortium, Tech. Rep. DOT HS 809 859, Mar. 2005.
5. W. Zhuang and H. A. Omar, "Vehicular communication netwroks: opportunities and challenges," in *Proc. 5th International Symposium and the 4th Student Organizing International Mini-Conference on Information Electronics Systems*, Feb. 2012.
6. R. Baldessari, B. Bödekker, A. Brakemeier, M. Deegener, A. Festag, W. Franz, A. Hiller, C. Kellum, T. Kosch, A. Kovacs, M. Lenardi, A. Lübke, C. Menig, T. Peichl, M. Roeckl, D. Seeberger, M. Strassberger, H. Stratil, H.-J. Vögel, B. Weyl, and W. Zhang, "Car-2-car communication consortium manifesto," Tech. Rep. Version 1.1, Aug. 2007.
7. "Specification for Telecommunications and Information Exchange Between Roadside and Vehicle Systems 5 GHz B and Dedicated Short Range Communications (DSRC) Medium Access Control (MAC) and Physical Layer (PHY) Specifications," *ASTM Standard E2213*, 2003 (2010).
8. http://www.car-to-car.org/.
9. http://www.cvisproject.org/.
10. http://www.sevecom.org/.
11. http://evita-project.org/.
12. http://www.com2react-project.org/.
13. http://www.ertico.com.
14. http://www.safespot-8220;eu.org/.
15. http://www.comesafety.org/http://www.comesafety.org/.
16. http://www.geonet-8220;project.eu/.
17. http://www.ict-8220;itetris.eu/index.htm.
18. "IEEE standard for information technology-telecommunications and information exchange between systems-local and metropolitan area networks-specific requirements—Part 11: Wireless LAN medium access control (MAC) and physical layer (PHY) specifications," *IEEE Std 802.11–2012 (Revision of IEEE Std 802.11-2007)*, pp. 1–2793, Mar. 2012.

19. T. H. Luan, X. Ling, and X. Shen, "MAC in motion: Impact of mobility on the MAC of drive-thru Internet," *IEEE Transactions on Mobile Computing*, vol. 11, no. 2, pp. 305–319, Feb. 2012.

20. J. J. Blum and A. Eskandarian, "A reliable link-layer protocol for robust and scalable interve-hicle communications," *IEEE Transactions on Intelligent Transportation Systems*, vol. 8, no. 1, pp. 4–13, Mar. 2007.

21. R. Mangharam, R. Rajkumar, M. Hamilton, P. Mudalige, and F. Bai, "Bounded-latency alerts in vehicular networks," in *Proc. Mobile Networking for Vehicular Environments (MOVE 2007)*, May 2007, pp. 55–60.

22. F. Watanabe, M. Fujii, M. Itami, and K. Itoh, "An analysis of incident information transmission performance using MCS/CDMA scheme," in *Proc. IEEE Intelligent Vehicles Symposium (IV'05)*, Jun. 2005, pp. 249–254.

23. H. Nakata, T. Inoue, M. Itami, and K. Itoh, "A study of inter vehicle communication scheme allocating PN codes to the location on the road," in *Proc. IEEE Intelligent Transportation Systems Conference (ITSC 2003)*, vol. 2, Oct. 2003, pp. 1527–1532.

24. F. Borgonovo, A. Capone, M. Cesana, and L. Fratta, "ADHOC MAC: New MAC architecture for ad hoc networks providing efficient and reliable point-to-point and broadcast services," *Wireless Networks*, vol. 10, pp. 359–366, Jul. 2004. [Online]. Available: http://dx.doi.org/10.1023/B:WINE.0000028540.96160.8a.

25. F. Borgonovo, L. Campelli, M. Cesana, and L. Coletti, "MAC for ad-hoc inter-vehicle network: services and performance," in *Proc. IEEE Vehicular Technology Conference (VTC 2003-Fall)*, vol. 5, Oct. 2003, pp. 2789–2793.

26. F. Borgonovo, L. Campelli, M. Cesana, and L. Fratta, "Impact of user mobility on the broadcast service efficiency of the ADHOC MAC protocol," in *Proc. IEEE Vehicular Technology Conference (VTC 2005-Spring)*, vol. 4, Jun. 2005, pp. 2310–2314.

27. S. Bharati and W. Zhuang, "CAH-MAC: Cooperative ADHOC MAC for vehicular networks," *IEEE Journal on Selected Areas in Communications*, vol. 31, no. 9, pp. 470–479, Sept. 2013.

28. S. Bharati, L. Thanayankizil, F. Bai, and W. Zhuang, "Effects of time slot reservation in cooperative ADHOC MAC for vehicular networks," in *Proc. IEEE International Conference on Communications (ICC)*, Jun. 2013, pp. 6371–6375.

29. S. Bharati and W. Zhuang, "Performance analysis of cooperative ADHOC MAC for vehicular networks," in *Proc. IEEE Global Communications Conference (GLOBECOM)*, Dec. 2012, pp. 5482–5487.

30. "IEEE standard for information technology–telecommunications and information exchange between systems–local and metropolitan area networks–specific requirements Part 11: Wireless LAN medium access control (MAC) and physical layer (PHY) specifications Amendment 6: Wireless access in vehicular environments," *IEEE Std 802.11p-2010 (Amendment to IEEE Std 802.11-2007 as amended by IEEE Std 802.11k-2008, IEEE Std 802.11r-2008, IEEE Std 802.11y-2008, IEEE Std 802.11n-2009, and IEEE Std 802.11w-2009)*, pp. 1–51, Jul. 15, 2010.

31. "IEEE standard for wireless access in vehicular environments (WAVE) - multi-channel operation," *IEEE Std 1609.4-2010 (Revision of IEEE Std 1609.4-2006)*, pp. 1–89, Feb. 2011.

32. H. A. Omar, W. Zhuang, and L. Li, "Evaluation of VeMAC for V2V and V2R Communications Under Unbalanced Vehicle Traffic," *Proc. IEEE Vehicular Technology Conference (VTC2012-Fall)*, Sept. 2012.

33. H. A. Omar, W. Zhuang, and L. Li, "VeMAC: a novel multichannel MAC protocol for vehicular ad hoc networks," in *Proc. IEEE International Conference on Computer Communications (INFOCOM) Workshop on Mobility Management in the Networks of the Future World (MobiWorld 2011)*, Apr. 2011, pp. 413–418.

Chapter 2
System Model

2.1 VANET Description: Elements and Applications

The VANET under consideration consists of a set of RSUs and a set of vehicles moving in opposite directions on two-way vehicle traffic roads. A vehicle is said to be moving in a left (right) direction if it is currently heading to any direction from north/south to west (east). Based on this definition, as shown in Fig. 2.1, if two vehicles are moving in opposite directions on a two-way road, it is guaranteed that one vehicle is moving in a left direction while the other vehicle is moving in a right one, regardless of the orientation of the road. The vehicles and RSUs broadcast safety messages (periodic and event-driven) for the purpose of safety applications. The periodic safety messages broadcast by different vehicles have the same (fixed) message size[1]. Also, the periodic safety messages broadcast by an RSU have equal message size, which may differ from the size of the periodic messages broadcast by another RSU based on the application.

2.2 Communications Channels

The VANET has one CCH and N_{sch} service channels (SCHs), denoted by $c_1, c_2, \ldots,$ $c_{N_{sch}}$. The CCH is used for transmission of two kinds of information: high priority short applications (such as periodic or event driven safety messages), and control information required for the nodes to organize the communications over the service channels. The N_{sch} SCHs are used for transmission of safety or non-safety related application messages. It is assumed that the transmission power levels on all channels are fixed and known to all nodes. All channels are symmetric, in the sense that node x

---

[1] A generic format of a periodic safety message, called the Basic Safety Message (BSM), is specified in the SAE J2735 application layer standard [1] to be broadcast by vehicles. The BSM makes use of the large similarity among the vehicle state information required by various V2V applications in order to avoid using a specific message for each application, which may result in a waste of the wireless network resources [2].

© The Author(s) 2014
H. A. Omar, W. Zhuang, *Time Division Multiple Access For Vehicular Communications*,
SpringerBriefs in Computer Science, DOI 10.1007/978-3-319-09504-2_2

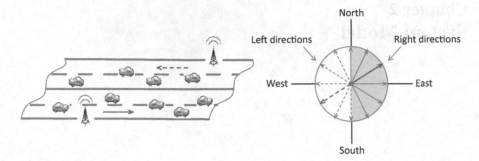

Fig. 2.1 Right and left directions of vehicle movement [3]

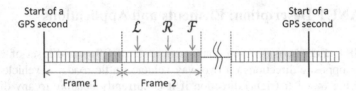

Fig. 2.2 Partitioning of each frame on the CCH into $\mathcal{L}, \mathcal{R}$ and $\mathcal{F}$ sets [3]

is in the communication range of node y if and only if node y is in the communication range of node x.

On the CCH, the time is partitioned to frames consisting of a constant number L of time slots of equal duration t. Each frame is partitioned into three sets of time slots: $\mathcal{L}, \mathcal{R}$, and $\mathcal{F}$, in that order, as shown in Fig. 2.2. The $\mathcal{F}$ set is associated with RSUs, while the $\mathcal{L}$ and $\mathcal{R}$ sets are associated with vehicles moving in left and right directions respectively. Each second contains an integer (fixed) number of frames, and each time slot is identified by the index (from 0 to $L-1$) of the time slot within a frame.

2.3   Node Equipment and Identification

Each node (i.e., vehicle or RSU) has two transceivers: Transceiver1 is always tuned to the CCH, while Transceiver2 switches among the SCHs. Also, each node is equipped with a global positioning system (GPS) receiver and can accurately determine its position and moving direction using GPS. The current position of each node is included in the header of each packet transmitted on the CCH. Each node is identified by a unique MAC address and a set of short identifiers (IDs), called VeMAC ID, where each VeMAC ID corresponds to a certain time slot that the node is accessing per frame on the CCH (more details in Chap. 3). Each VeMAC ID is chosen by a node at random, included in the header of each packet transmitted in the corresponding time slot, and changed if the node detects that its ID is already in use by another node [4].

2.4 Time Slot Synchronization

Synchronization among nodes is performed using the one pulse per second (1PPS) signal provided by any GPS receiver. The rising edge of this 1PPS is aligned with the start of every GPS second with accuracy within 100 ns even for inexpensive GPS receivers. Hence, this 1PPS signal can be used as an accurate common time reference among all the nodes. Consequently, at any instant, each node can determine the index of the current slot within a frame on the CCH, and whether it belongs to the $\mathcal{L}$, $\mathcal{R}$, or $\mathcal{F}$ set. In case of a temporary loss of GPS signal, the synchronization among different nodes can still be maintained within a certain accuracy for a time duration which depends mainly on the stability of the local oscillator of the GPS receiver at each node [5]. If the GPS signal is lost in a certain area for a duration longer than a specified threshold, a distributed synchronization scheme, e.g., [5], should be employed until the GPS signal is recovered. Details of such a back up distributed synchronization scheme are out of scope of this brief.

2.5 Definitions

For a certain node, x, set $\mathcal{N}_x$ denotes the set of one-hop neighbours of node x, from which node x has received VeMAC Type1 packets (defined in Sect. 3.1.3) on the CCH in the previous L slots. Set $\mathcal{T}_x$ is defined as the set of time slots that node x must not use on the CCH in the next L time slots. This set is used by node x to determine which time slots it can access on the CCH without causing any hidden terminal problem. How each node x constructs and updates sets $\mathcal{N}_x$ and $\mathcal{T}_x$ is discussed in Chap. 3.

A two-hop set (THS) is defined as a set of nodes in which each node can reach any other node in two hops at most. The term 'packet' refers to a MAC layer protocol data unit (MPDU), and the term 'message' refers to a MAC layer service data unit (MSDU), i.e., the unit of information arriving to the MAC layer entity from the layer above.

References

1. "Dedicated short range communications (DSRC) message set dictionary," *SAE J2735 Standard*, Nov. 19, 2009.
2. F. Zaid, F. Bai, S. Bai, C. Basnayake, B. Bellur, S. Brovold, G. Brown, L. Caminiti, D. Cunningham, H. Elzein, J. Ivan, D. Jiang, J. Kenny, H. Krishnan, J. Lovell, M. Maile, D. Masselink, E. McGlohon, P. Mudalige, V. Rai, J. Stinnett, L. Tellis, K. Tirey, and S. VanSickle, "Vehicle safety communications-applications (VSC-A) second annual report," The CAMP Vehicle Safety Communications 2 Consortium, Tech. Rep. DOT HS 811 466, Aug. 2011.
3. H. A. Omar, W. Zhuang, A. Abdrabou, and L. Li, "Performance evaluation of VeMAC supporting safety applications in vehicular networks," *IEEE Transactions on Emerging Topics in Computing*, vol. 1, no. 1, pp. 69–83, Jun. 2013.

4. F. Borgonovo, A. Capone, M. Cesana, and L. Fratta, "ADHOC MAC: New MAC architecture for ad hoc networks providing efficient and reliable point-to-point and broadcast services," *Wireless Networks*, vol. 10, pp. 359–366, Jul. 2004. [Online]. Available: http://dx.doi.org/10.1023/B:WINE.0000028540.96160.8a.

5. K. Kutzner, J.-J. Tchouto, M. Bechler, L. Wolf, B. Bochow, and T. Luckenbach, "Connecting vehicle scatternets by Internet-connected gateways," in *Proc. Workshop on Multiradio Multimedia Communications (MMC 2003)*, University of Dortmund, Germany, 2003. [Online]. Available: http://i30www.ira.uka.de/research/publications/p2p/.

Chapter 3
The VeMAC Protocol

This chapter presents VeMAC, a novel multichannel TDMA protocol developed specifically for a VANET scenario [1, 2, 3]. The VeMAC supports an efficient one-hop broadcast services on the CCH, by using implicit acknowledgments and eliminating the hidden terminal problem, in order to successfully deliver both periodic and event-driven safety messages in VANETs. The protocol reduces transmission collisions due to node mobility on the control channel by assigning disjoint sets of time slots to vehicles moving in opposite directions and to road side units. Analysis and simulation results in highway and city scenarios are presented to evaluate the performance of VeMAC and compare it with ADHOC MAC [4]. It is shown that, due to its ability to reduce the rate of transmission collisions, the VeMAC protocol can provide considerably higher throughput on the CCH than ADHOC MAC.

3.1 VeMAC Basics

3.1.1 Safety Message Queueing and Service

As discussed in Sect. 2.1, the vehicles and RSUs broadcast periodic and event-driven safety messages on the CCH for the purpose of safety applications, and the periodic safety messages broadcast by different vehicles have the same (fixed) message size. As shown in Fig. 3.1, at each node, the periodic and event-driven safety messages are mapped to two different queues, which are served independently by the VeMAC protocol, as described in details in Sect. 3.2.

Based on a given transmission rate determined by the physical layer, the VeMAC maximum transmission unit (MTU) is defined as the maximum amount of data (without the physical layer overhead) which can be transmitted in the duration of one time slot. The duration of a time slot, t, is chosen such that the MTU is equal to the size of a periodic safety message broadcast by a vehicle plus the maximum size of control information introduced by the VeMAC protocol. For RSUs, if the size of a periodic safety message plus the VeMAC control information exceeds the MTU, the message is fragmented to be transmitted as multiple VeMAC packets, as indicated in

© The Author(s) 2014
H. A. Omar, W. Zhuang, *Time Division Multiple Access For Vehicular Communications*,
SpringerBriefs in Computer Science, DOI 10.1007/978-3-319-09504-2_3

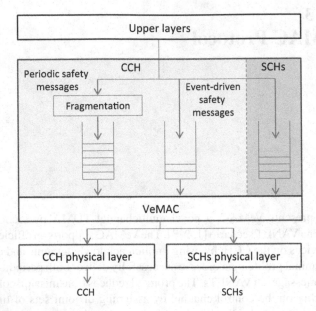

Fig. 3.1 Safety message queues [2]

Fig. 3.1. This fragmentation is typical for applications such as curve speed warning and left turn assistant [5], in which the size of a periodic safety message broadcast by an RSU is considerably larger than that of the periodic messages broadcast by vehicles [5]. On the other hand, all the event-driven safety messages are assumed to be small enough to fit in a single VeMAC packet, without fragmentation. Each VeMAC packet carries at most one safety message and only one VeMAC packet can be transmitted per time slot.

In the VeMAC protocol, in order to serve the two safety message queues in Fig. 3.1, each node must acquire at least one time slot per frame on the CCH. A time slot acquired by a certain node is referred to as a periodic or event-driven slot, according to the type of the safety message transmitted during this time slot. The number of periodic slots that the node acquires per frame, denoted by k_p, depends on the fixed size and arrival rate of the periodic safety messages. Similarly, the number of event-driven slots that the node can access per frame, denoted by k_e, depends on the average arrival rate of the event-driven safety messages. A node should use a unique VeMAC ID to access each of the k_p and k_e slots. As mentioned in Sect. 2.3, each VeMAC ID is chosen by the node at random, included in the header of each packet transmitted in the corresponding time slot, and changed if the node detects that its VeMAC ID is already in use by another node [4]. The k_p and k_e values are chosen such as to satisfy the delay constraints of the periodic and event-driven safety messages based on the delay analysis in Chap. 4. Once a node acquires a periodic or event-driven slot, it keeps using the same slot in all subsequent frames unless there is no packet waiting for transmission in the corresponding queue or a transmission collision is detected.

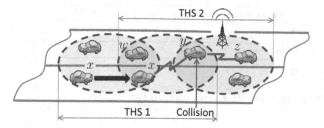

Fig. 3.2 Merging collision caused by node mobility [1]

3.1.2 Transmission Collision Types on The CCH

Two types of transmission collision can happen on the CCH: access collision and merging collision. An access collision happens when two or more members of the same THS attempt to acquire the same available time slot. On the other hand, a merging collision happens when two or more nodes acquiring the same time slot become members of the same THS due to node activation or node mobility. In VANETs, merging collisions are more likely to occur among vehicles moving in opposite directions or between a vehicle and a stationary RSU since they approach each other with a much higher relative velocity as compared to vehicles moving in the same direction. For example, in Fig. 3.2, if vehicle x moves to THS2 and if x is using the same time slot as z, then collision will occur at y. Upon detection of a merging collision on the CCH, each colliding node should release its time slot and acquire a new one, which may generate more access collisions.

3.1.3 VeMAC Packet Types

Two different types of VeMAC packets can be transmitted on the CCH, as shown in Fig. 3.3. A Type1 packet is divided into four main fields: Type1 header, announcement of services (*AnS*), acceptance of services (*AcS*), and high priority short applications (*HPSA*). The *HPSA* field is to include the periodic and event-driven safety messages, while the *AnS* and *AcS* fields are used to control the communications over the SCHs. A Type2 packet does not contain any control information: it consists of an *HPSA* field and a short Type2 header (the difference between Type1 and Type2 headers will be discussed). Each node must transmit exactly one Type1 packet in each frame using one of its acquired periodic time slots, and if the node is accessing more than one time slot per frame, Type2 packets are transmitted over the rest of time slots. The transmission of one Type1 packet in each frame is mandatory since the information in the Type1 header, *AnS* and *AcS* fields, is necessary for other nodes to decide which time slots they can access on the CCH and SCHs. On the other hand, the transmission of Type2 packets is to decrease the protocol overhead by removing all the control information which needs to be transmitted only once per frame. As the event-driven safety messages are always transmitted using Type2 packets

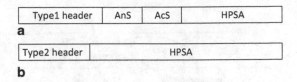

Type1 header	AnS	AcS	HPSA

a

Type2 header	HPSA

b

Fig. 3.3 VeMAC packet types [2]. **a** Type1 packet.**b** Type2 packet

(i.e., without control information and with a large *HPSA* field in the packet), fragmentation is not considered for this type of safety messages.

3.2   CCH Access

For the purpose of time slot assignment on the CCH, in the header of each Type1 packet transmitted on the CCH, the transmitting node x should broadcast the VeMAC ID(s) and the corresponding time slot(s) of each node in set $\mathcal{N}_x$. The short length of a VeMAC ID (9 bits as suggested in Chap. 4) serves to decrease the protocol overhead as compared to broadcasting the corresponding MAC address. The main difference between Type1 and Type2 headers is that the Type2 one is shorter as it does not contain the VeMAC IDs or the corresponding time slots of the nodes in set $\mathcal{N}_x$. Now, suppose node x is just powered on and needs to acquire a time slot. By listening to the CCH for L successive time slots (not necessarily in the same frame), node x can determine set $\mathcal{N}_x$ and the time slot(s) used by each node in $\mathcal{N}_x$. Also, since each one-hop neighbour $y \in \mathcal{N}_x$ announces (in the header of its transmitted Type1 packet) the time slot(s) used by each node in $\mathcal{N}_y$, node x can determine all the time slots used by each of its two-hop neighbours, $z \in \mathcal{N}_y, z \notin \mathcal{N}_x, \forall y \in \mathcal{N}_x$. Accordingly, node x sets $\mathcal{T}_x$ to the set of time slots used by all nodes within its two-hop neighbourhood. Then, sets $\mathcal{N}_x$ and $\mathcal{T}_x$ are updated by node x at the end of each time slot (always based on the packets received in the previous L slots).

Given $\mathcal{T}_x$, node x determines the set of accessible time slots, $\mathcal{A}_x$, (to be discussed) and then attempts to acquire a time slot by randomly accessing any time slot in $\mathcal{A}_x$, say time slot k. If no other node in the two-hop neighbourhood of node x simultaneously attempts to acquire time slot k, then no access collision happens. In this case, the attempt of node x is successful and each one-hop neighbour w of node x adds node x to set $\mathcal{N}_w$ and record the VeMAC ID used by node x to access time slot k, denoted by ID_k^x. On the other hand, if at least one node within the two-hop neighbourhood of node x accesses time slot k, then all the transmissions in the slot fail and time slot k is not acquired by any of the contending nodes. Node x will determine whether or not its attempt was successful by observing the $L - 1$ time slots following k. The attempt of node x is considered successful iff the Type1 packet received from each node $w \in \mathcal{N}_x$ includes ID_k^x in its headers. Otherwise, node x re-accesses one of the time slots in $\mathcal{A}_x$ until it successfully acquires a time slot. Once node x acquires a time slot, it keeps using the same slot in all subsequent frames unless a merging collision happens. Similar to an access collision, a merging collision on time slot k is

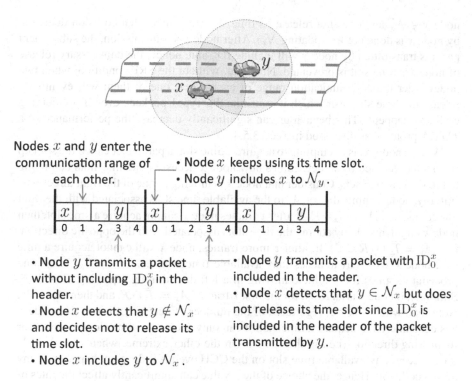

Fig. 3.4 The SRP condition preventing node x from unnecessarily releasing its time slot

detected by node x as soon as it receives a Type1 packet from a node $w \in \mathcal{N}_x$ without including ID_k^x in its header. Upon detection of a merging collision, each colliding node should release its time slot and acquire a new one using the same procedure. In order to acquire more than one time slot per frame, node x employs the same procedure using a unique VeMAC ID for accessing each extra time slot.

At the end of each time slot, the collision detection by a certain node x should be done *before* updating set $\mathcal{N}_x$. Upon receiving a Type1 packet from a node y without including ID_k^x in its header, we stress on that, node x should approve this collision detection and release time slot k iff the transmitting node $y \in \mathcal{N}_x$. This condition is referred to as the slot release prevention (SRP) condition, and its main objective is to prevent node x from unnecessarily releasing its time slot when it just enters the communication range of another node y. To illustrate that, consider the time slot assignment shown in Fig. 3.4 for the two nodes x and y. Note that, each of the nodes x and y is accessing one time slot per frame, and hence all the packets transmitted are of Type1 packets. When node x enters the communication range of node y, even if no collision happens, the first packet received by node x from node y will not include ID_0^x. The reason is that, by the time node y transmits its packet, node y has not yet received any packet from node x on time slot 0. By applying the SRP condition, when node x receives the first packet from node y, node x determines that node $y \notin \mathcal{N}_x$ and does not release its time slot (remember that collision detection by node x is done before updating $\mathcal{N}_x$). After node x's transmission, the subsequent packets transmitted by node y will include ID_0^x and, hence, the unnecessary release of node x's time slot is prevented. Note that, without the SRP condition, when two nodes enter the communication range of each other, one of them will eventually release the time slot over which it transmits the Type1 packets, even if no merging collision happens. This behaviour can significantly decrease the performance of a TDMA protocol as discussed in Sect. 3.5.4.

When a node, x, is attempting to acquire a time slot, a parameter called the split up parameter, denoted by τ, determines how node x accesses the time slots belonging to the $\mathcal{L}$, $\mathcal{R}$, and $\mathcal{F}$ sets. Consider that node x is moving in one of the right directions. Initially, node x limits the set $\mathcal{A}_x$ to the available time slots associated with the right directions, i.e., $\mathcal{A}_x = \overline{\mathcal{T}_x} \cap \mathcal{R}$. If after τ frames node x cannot acquire a time slot, then node x augments $\mathcal{A}_x$ by adding the time slots associated with the opposite direction, i.e., $\mathcal{A}_x = \overline{\mathcal{T}_x} \cap (\mathcal{R} \cup \mathcal{L})$. If, after τ more frames, node x still cannot acquire a time slot, node x will start to access any available time slot, i.e., $\mathcal{A}_x = \overline{\mathcal{T}_x}$. The same procedure applies for a vehicle moving in a left direction by replacing $\mathcal{R}$ with $\mathcal{L}$. Similarly, if node x is an RSU, for the first τ frames $\mathcal{A}_x = \overline{\mathcal{T}_x} \cap \mathcal{F}$, and then $\mathcal{A}_x = \overline{\mathcal{T}_x}$. Note that, when $\tau = \infty$, regardless of the number of access collisions that node x has encountered to acquire a time slot, it can only access the time slots reserved for its moving direction (i.e., in the $\mathcal{R}$ set). On the other extreme, when $\tau = 0$, node x can access any available time slot on the CCH even if it does not experience any access collision. Hence, the choice of the τ value can significantly affect the rates of access collision and merging collision. For example, when $\tau = 0$, all the vehicles and RSUs are accessing the same set of time slots. Hence, a merging collision is possible between any two nodes. However, when a merging collision happens, each colliding node x is free to access any time slot in $\overline{\mathcal{T}_x}$, which can decrease the probability of an access collision. On the other extreme, when $\tau = \infty$, the vehicles moving in opposite directions and the RSUs are accessing disjoint sets of time slots. However, when a merging collision happens, for example among vehicles moving in a right direction, there is a higher probability of an access collision (compared with the $\tau = 0$ case) since the choice of each colliding node x is limited to time slots in $\overline{\mathcal{T}_x} \cap \mathcal{R}$. A performance comparison between the $\tau = 0$ and $\tau = \infty$ cases is provided in Sect. 3.5, and the effect of the τ value on the delay of periodic and event-driven safety messages is investigated in Chap. 4 for these two extreme cases.

Using the proposed scheme, a reliable broadcast service can be provided on the CCH. That is, if node x transmits a broadcast packet on time slot k, by listening to the $L - 1$ time slots following k, node x can determine the set $\mathcal{D}_x$ of one-hop neighbors which have not successfully received the packet, where $\mathcal{D}_x = \{y \in \mathcal{N}_x :$ ID_k^x is not broadcast by node $y\}$. In other words, when node y includes ID_k^x in the header of its Type1 packet, it is considered as an implicit acknowledgement by node y of receiving the packet broadcast by node x on time slot k.

3.3 SCH Access

On the SCHs, time is partitioned to frames consisting of a constant number of fixed duration time slots. All the SCHs are slot synchronized with the CCH, and on each SCH, each second contains an integer number of frames. The number of time slots per frame on channel c_m is denoted by l_m, $m = 0, \ldots, N_{sch}$, and a time slot on channel c_m is identified by the index (from 0 to $l_m - 1$) of this time slot within a frame on channel c_m. Note that, the same time slot can have different indices on channels c_i and c_j, $i \neq j$, since l_i is not necessarily equal to l_j.

A *provider* is a node which announces on the CCH for a service offered on a specific SCH, while a *user* is a node which receives the announcement for a service and decides to make use of this service[1]. For a certain node x, let T_x^m denote the set of time slots that node x must not use on channel c_m in the next l_m time slots, $m = 0, \ldots, N_{sch}$. Set T_x^m is used by node x to determine which time slots it can access on channel c_m without causing any hidden terminal problem, as described in the following.

Consider that a node x has an MSDU to be delivered to a certain destination (assuming unicast) on SCH c_m. By using T_x^m (how node x constructs T_x^m will be explained), node x determines the set of time slots that it will access on channel c_m to offer the service, denoted by ϑ_x^m, such that $\vartheta_x^m \cap T_x^m = \phi$. Accordingly, node x announces the following information in the AnS field of its next packet transmitted on the CCH: (a) priority of the service, (b) reliability of the service (i.e., acknowledged or not), (c) MAC address of the intended destination y, (d) the index m of the service channel, and (e) ϑ_x^m. Once the provider x announces for the service, no further action is needed unless the destination accepts the service as described below.

Based on the information announced by provider x on the CCH, the destination y determines whether or not to make use of the announced service. If node y decides to use the service by provider x on channel c_m, it accepts the service by including ϑ_x^m in the AcS field of its next packet transmitted on the CCH. The announcement of ϑ_x^m by the user y is for each surrounding node, z, to update its T_z^m set as to be discussed. Also, for a reliable service, node y should include in the AnS field the time slot that will be used by node y to transmit the acknowledgement packet, denoted by a_y^m. Node y determines a_y^m such that $a_y^m \notin T_y^m$. When provider x receives the acceptance of the service, it tunes its Transceiver2 to channel c_m and starts offering the service on the time slots announced in ϑ_x^m. As well, if the service is reliable, node x should include a_y^m in the AcS field of its next packet transmitted on the CCH. Again, the announcement of a_y^m by provider x is to avoid the collision of the acknowledgement packet by properly updating the T_z^m set of each surrounding node z. Node y should transmit the acknowledgement only after node x announces a_y^m on the CCH.

Each node, x, updates sets T_x^m, $m = 1, \ldots, N_{sch}$, as follows. When node x receives a packet on the CCH from another node y, based on the position of node y which is included in the header of the packet, and the position of node x obtained from the GPS receiver, node x can estimate its distance to node y. Based on this estimated

---

[1] The term 'service' refers to the delivery of an MSDU on a certain service channel.

distance and on the fixed transmission power on all channels which is known to node x, node x can determine whether or not node y is in its communication range on channel c_m, $m = 1, \ldots, N_{sch}$ [2]. If node x decides that it can reach node y on a certain channel c_m, node x adds to set $\mathcal{T}_x^m$ the time slots indicated by each ϑ_y^m set and a_y^m slot included in the AcS field of the packet transmitted by y. The reason is that, each ϑ_y^m represents a set of time slots over which node y will receive a packet on channel c_m from a certain provider in the next l_m slots. Similarly, each a_y^m indicates a time slot over which node y will receive an acknowledgement packet on channel c_m from a certain user in the next l_m slots. Consequently, by updating $\mathcal{T}_x^m$ in the way described, collision at node y can be prevented since each node x in the one-hop neighbourhood of node y will avoid using the time slots over which node y will receive packets. At the end of time slot i_m on channel c_m, if $i_m \in \mathcal{T}_x^m$, node x removes i_m from $\mathcal{T}_x^m$, $m = 1, \ldots, N_{sch}$ and $i_m = 0, \ldots, l_m - 1$. Note that updating the $\mathcal{T}_x^m$, $m = 1 \ldots, N_{sch}$, sets for each node x is based on information in the AcS (not in the AnS) field, which eliminates any exposed terminal problem. The following example illustrates how the nodes access the service channels.

Consider the THS configuration shown in Fig. 3.5, node x has a reliable service to offer to node z on time slots numbered 1, 2, and 4 on channel c_1. Figure 3.5 shows the sequence of actions taken by provider x, user z, and the surrounding nodes y and w. First, node x announces for the service and includes $\vartheta_x^1 = \{1, 2, 4\}$ in the AnS field of its packet transmitted on the CCH. Following this announcement, no action is taken by both surrounding nodes w and y. Once node z accepts the service and announces ϑ_x^1, node x starts offering the service on channel c_1 on time slots $\{1, 2, 4\}$ as announced in ϑ_x^1. When node y receives the packet transmitted by node z on the CCH, it adds ϑ_x^1 to $\mathcal{T}_y^1$ to avoid using the upcoming time slots $\{1, 2, 4\}$ over which node z will receive packets from node x (assume that node y can reach node z on channel c_1). Note that, node w is free to use the time slots in $\vartheta_x^1 = \{1, 2, 4\}$ since it did not receive the acceptance of service transmitted by node z on the CCH; hence, simultaneous transmissions from node w to v and from node x to z are allowed on channel c_1, i.e., no exposed terminal problem. However, in the absence of the exposed terminal problem, it is possible that node w announces a service to node y on time slots $\{1, 2, 4\}$ after node x did the same announcement to node z (note that simultaneous transmissions from node w to y and from node x to z result in a collision at node y). In this case, if node y accepts the service and includes $\vartheta_w^1 = \{1, 2, 4\}$ in the AcS field of its packet transmitted on the CCH (on time slot $\{6\}$), node x will receive this packet transmitted by node y, includes ϑ_w^1 to $\mathcal{T}_x^1$, and avoids using the upcoming time slots $\{1, 2, 4\}$ on channel c_1 to prevent collision at node y (recall the definition of $\mathcal{T}_x^1$), although node x was supposed to transmit a packet to node z on the time slot $\{2\}$ following node y's acceptance of service. This missing packet, together with the other packets incorrectly received by node z, are (re)transmitted by node x after it receives the acknowledgment packet from node z. The acknowledgement packet is transmitted using the same procedure as illustrated in Fig. 3.5.

---

[2] It is assumed that each node has a path loss model for each service channel c_m, $m = 1, \ldots, N_{sch}$.

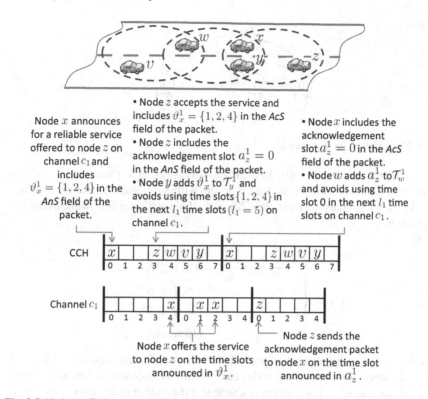

Fig. 3.5 Node x offering a service to node z on channel c_1

3.4   Analysis of Time Slot Acquisition

The objective of the analysis in this subsection is to investigate how fast the contending nodes can acquire a time slot on the CCH by using the VeMAC protocol. Let K denote the number of contending nodes, each of which needs to acquire a time slot on the CCH. We want to determine the average number of nodes which acquire a time slot within n frames, the probability that a specific node acquires a time slot within n frames, and the probability that all the nodes acquire a time slot within n frames. To simplify the analysis, the following assumptions are made: (a) all the contending nodes belong to the same set of THSs, e.g., node w and node x in its final position in Fig. 3.2; (b) the set of THSs to which the contending nodes belong does not change; (c) when a node, x, fails to acquire a time slot after τ frames, the set $\mathcal{A}_x$ is not augmented, i.e., $\tau = 0$; (d) at the end of each frame, each node, x, is aware of all acquired time slots during the frame, and updates the sets $\mathcal{T}_x$ and $\mathcal{A}_x$ accordingly, i.e., all nodes are within the communication range of each other; (e) at the end of each frame, all contending nodes are informed whether or not their attempts to access a time slot during this frame were successful. Based on this information, each colliding node, x, randomly chooses an available time slot from the updated $\mathcal{A}_x$ set, and attempts to access this slot during the coming frame.

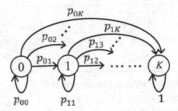

Fig. 3.6 Markov chain for X_n when $K \leq N$ [1]

Let N be the number of initially available time slots in a frame, and X_n be the total number of nodes which acquired a time slot within n frames. Under the assumptions, X_n is a stationary discrete-time Markov chain with the following transition probabilities.

If $K \leq N$,

$$p_{ij} = \begin{cases} \frac{W(j-i, K-i, N-i)}{(N-i)^{K-i}}, & 0 \leq i \leq K-1, \\ & i \leq j \leq K \\ 1, & i = j = K \\ 0, & \text{elsewhere} \end{cases}$$

where $W(l, u, v)$ is the number of ways by which l nodes can acquire a time slot given that there are u contending nodes each randomly choosing a time slot among v available time slots. A node acquires a time slot if no other nodes choose to access the same slot. The Markov chain is illustrated in Fig. 3.6.

If $K > N$,

$$p_{ij} = \begin{cases} \frac{W(j-i, K-i, N-i)}{(N-i)^{K-i}}, & 0 \leq i \leq N-1, \\ & i \leq j \leq N-1 \\ 1, & i = j, N \leq i \leq K \\ 0, & \text{elsewhere.} \end{cases}$$

The Markov chain is illustrated in Fig. 3.7. To calculate $W(l, u, v)$, considering u different balls randomly distributed in v different boxes with equal probabilities, $W(l, u, v)$ is the number of ways of having l boxes each containing exactly one ball. This special occupancy problem is solved in a recursive way as follows [6].

If $u \leq v$,

$$W(l, u, v) = \begin{cases} C_l^u A_l^v \Big((v-l)^{u-l} - \\ \sum_{i=1}^{u-l} W(i, u-l, v-l) \Big), & 0 \leq l < u \\ A_l^v, & l = u \\ 0, & l > u \end{cases}$$

where $A_l^v = \frac{v!}{(v-l)!}$ and $C_l^u = \frac{A_l^u}{l!}$.

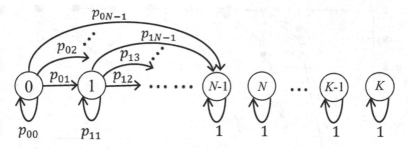

Fig. 3.7 Markov chain for X_n when $K > N$ [1]

If $u > v$,

$$W(l, u, v) = \begin{cases} C_l^u A_l^v \Big((v-l)^{u-l} - \\ \quad \sum_{i=1}^{v-l} W(i, u-l, v-l) \Big), & 0 \le l < v \\ 0, & l \ge v. \end{cases}$$

Let P be the one-step transition probability matrix, and P^n the n-step transition probability matrix. Given that initially all nodes are contending for time slots, i.e., $X_0 = 0$ with probability 1, the unconditional probability distribution of X_n is represented by the first row of P^n. That is,

$$p(X_n = i) = P_{1,i+1}^n, i = 0, \dots, K$$

where $P_{1,i+1}^n$, denotes the entry of the matrix P^n, located at the first row and $(i+1)^{\text{st}}$ column. The probability that all nodes acquire a time slot within n frames is denoted by F_n^{all}, where

$$F_n^{all} = p(X_n = K) = P_{1,K+1}^n.$$

Let μ_n denote the average number of nodes which acquire a time slot within n frames. Therefore,

$$\mu_n = \sum_{i=0}^{K} i P_{1,i+1}^n.$$

The probability that a specific node, say node x, acquires a time slot within n frames is denoted by F_n, where

$$F_n = \sum_{i=0}^{K} p(\mathcal{E} | X_n = i) p(X_n = i)$$

$$= \sum_{i=1}^{K} \frac{C_{i-1}^{K-1}}{C_i^K} P_{1,i+1}^n = \frac{\mu_n}{K}.$$

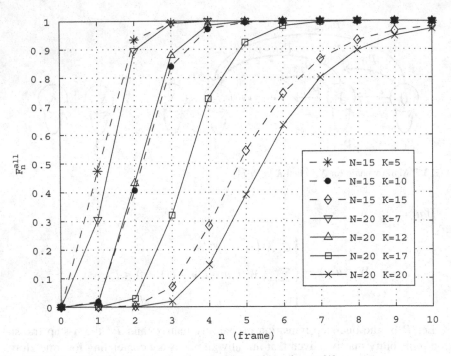

Fig. 3.8 Probability that all nodes acquire a time slot within n frames [1]

where $\mathcal{E}$ is the event that node x acquires a time slot within n frames and $p(\mathcal{E}|X_n = i)$ $= \frac{C_{i-1}^{K-1}}{C_i^K} = \frac{i}{K}$ since all nodes have equal chances of acquiring a time slot. Note that, since the VeMAC assumes a fixed number of constant duration time slots in a frame on the CCH, the choice of the L value should always ensure that $K \leq N$. However, the analysis of the protocol for the case $K > N$ can be useful in order to determine an optimal value for L. This analysis gives an indication of how the protocol will behave if the number of nodes in a THS becomes larger than L.

Figure 3.8 illustrates F_n^{all} for different values of N and K. As shown in Fig. 3.8, in a dense scenario such as $(N = 15, K = 15)$, there is a probability greater than 0.9 that all the contending nodes acquire a time slot within 8 frames. Hence, given a frame duration around 100 ms (as discussed in Chap. 4), the simplifying assumption of invariant THSs (assumption b) is acceptable, since it is reasonable to assume that the THS configuration remains constant for a sufficiently large time after all the contending nodes acquire a time slot. The analysis presented in this subsection is verified in Sect. 3.5.1 via MATLAB simulations.

3.5 Simulations

This section presents MATLAB simulation results to study the accuracy of the analysis in Sect. 3.4, and to evaluate the performance of VeMAC as compared with ADHOC MAC in accessing the CCH in highway and city scenarios.

3.5.1 Analysis Verification

Simulations have been conducted using MATLAB to verify the analysis in Sect. 3.4. In the simulations, assumption e) is removed, and the average number of nodes which acquire a time slot within n frames is calculated for different K and N, and denoted by μ_n^{sim}. The 98 % confidence interval of μ_n^{sim} is less than 0.33 node for all n, K, and N. As shown in Fig. 3.9, the results of μ_n^{sim} obtained from simulations without assumption e) are very close to μ_n obtained from analysis for different K and N.

3.5.2 Simulation Scenarios and Performance Metrics

The first scenario under consideration is a segment of a two-way vehicle traffic highway. A vehicle can communicate with all the vehicles within its communication range, i.e., no obstacles. Each vehicle moves with a constant speed drawn from a normal distribution, and the number of vehicles on the highway segment remains constant during the simulation time. When a vehicle reaches one end of the highway segment, it re-enters the segment from the other end. For this reason, to prevent the unrealistic merging collisions caused by vehicles which jump from one end to the other end, if a vehicle is located at a distance $d \leq R$ from one end of the highway segment, where R denotes the communication range, it can communicate with vehicles located within a distance $R - d$ from the other end of the segment. In this way, for each traffic direction, the vehicles at the end of the segment act as if they are following the vehicles at the start of the segment.

The second scenario is a city grid layout consisting of three horizontal and three vertical two-way vehicle traffic streets. All the streets have the same dimensions, and the horizontal and vertical streets are evenly spaced resulting in four identical square city blocks (a city block is the smallest area that is surrounded by streets). The area of intersection of a horizontal street with a vertical one is referred to as a junction area. Each vehicle moves with a constant speed drawn from a normal distribution. When a vehicle reaches a junction area, it chooses one of all possible moving directions with equal probability (vehicles are not allowed to leave the simulation area during the simulation time). A vehicle located at a junction area can communicate with vehicles within its communication range located on both streets intersecting at the junction area. On the other hand, a vehicle located at a street but not at a junction

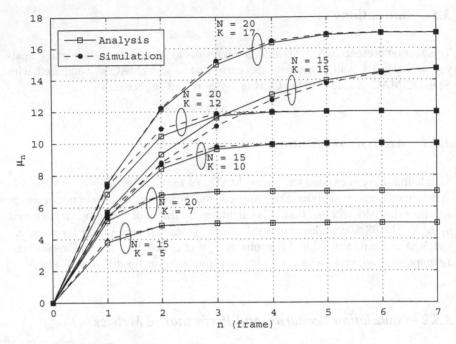

Fig. 3.9 Average number of nodes acquiring a time slot within n frames

area cannot communicate with vehicles located on other streets due to the existence of city blocks which obstruct the wireless signal.

For both scenarios under consideration, all the transmitted packets are broadcast packets, the wireless channel is ideal, and the only source of packet errors is the transmission collision. Table 3.1 summarizes the simulation parameters and Figs. 3.10 and 3.11 show snap shots of the simulated highway and city scenarios respectively, where the black and white triangles represent vehicles moving in opposite directions. The same slot duration and total number of time slots per frame are used for all the MAC protocols under consideration (Sect. 3.5.3).

We define a parameter, called the two-hop set occupancy (THSO), equal to $N_v \times \frac{2R}{\text{Length}_h} \times \frac{1}{L}$ or $\frac{N_v}{N_s} \times \frac{2R}{\text{Length}_s} \times \frac{1}{L}$ in the highway and city scenarios respectively, where N_v is the total number of vehicles, N_s is the total number of streets in the city, Length_h is the length of the highway segment, and Length_s is the length of a city street. Note that, the ratio $\frac{N_v}{N_s}$ approximately equals the number of vehicles on a city street, the number L represents the maximum number of time slots available for a THS, and the length $2R$ is the maximum length that a THS can occupy on the highway segment or on a city street. Consequently, the THSO indicates the ratio of the number of time slots required by a THS to the total number of time slots available for a THS. However, the THSO is not guaranteed for each THS in the simulations. The reason is that, if there are N_v moving vehicles, this does not mean that at each instant, each THS on the highway consists of $N_v \times \frac{2R}{\text{Length}_h}$ vehicles or each THS in the city consists

Table 3.1 Simulation parameters [1]

Parameter	Highway	City
Highway length	1 km	–
# horizontal streets	–	3
# vertical streets	–	3
City street length	–	430 m
# city blocks	–	4
City block edge length	–	200 m
# lanes/direction	2	1
Lane width	5 m	5 m
Speed mean value	100 km/h	50 km/h
Speed standard deviation	20 km/h	10 km/h
Transmission range	150 m	150 m
# slots/frame	100	100
# slots for left directions	50	50
# slots for right directions	50	50
# slots for RSUs	0	0
Slot duration	1 ms	1 ms
Simulation time	2 min.	2 min.
# vehicles	80 to 280 (step = 20)	150 to 600 (step = 50)
THSO	0.24 to 0.96 (step = 0.06)	0.17 to 0.70 (step = 0.06)

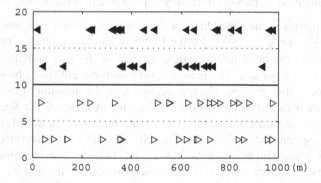

Fig. 3.10 A snap shotof the simulated highway segment [1]

of $\frac{N_y}{N_s} \times \frac{2R}{\text{Length}_s}$ vehicles. Also, in the city scenario under consideration, a THS located near a junction area can occupy a length on the streets up to $4R$ ($2R$ on each of the horizontal and vertical street intersecting at the junction area).

The following performance metrics are considered:

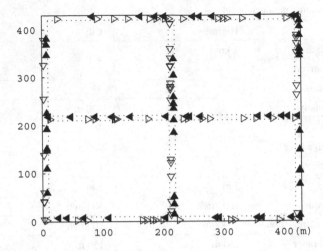

Fig. 3.11 A snap shot of the simulated area of the city [1]

a) rate of merging collisions: the average number of merging collisions per frame per THS;
b) rate of access collisions: the average number of access collisions per slot per THS;
c) Tx throughput: the average number of successful transmissions per slot per THS. A transmission by a vehicle x in a certain time slot is considered successful iff no other vehicles in the two-hop neighbourhood of x transmits in the same slot;
d) Rx throughput: the average number of successfully received packets per slot per THS. As mentioned, packet errors only happen due to transmission collision.

Each of the metrics is calculated first for the whole simulation area, and then multiplied by $\frac{2R}{\text{Length}_h}$ or $\frac{1}{N_s} \times \frac{2R}{\text{Length}_s}$ for the highway and city scenarios respectively. Note that, unlike the other three metrics, the rate of merging collisions is calculated per frame not per slot. The reason is that, merging collisions happen due to the movement of the vehicles, which is negligible in the duration of one time slot. The metrics are obtained for each of the MAC protocols mentioned in Sect. 3.5.3. At the beginning of the simulations, the vehicles are randomly (uniformly) placed on the highway segment and on all streets of the city. The vehicles remain stationary and try to acquire a time slot by using the MAC protocol under consideration. Once no more vehicle can acquire a time slot, the vehicles begin moving and the simulation timer starts. The objective of this process is to quickly bring the system to a steady state where most of the vehicles have acquired a time slot.

3.5.3   Simulated Protocols

Two versions of the VeMAC protocol are considered: VeMAC with $\tau = 0$ (V0) and VeMAC with $\tau = \infty$ (V-inf). As will be shown in Sect. 3.5.4, both versions of the VeMAC protocol significantly outperform the ADHOC MAC protocol in [4]. The poor performance of ADHOC MAC is caused by the following two main reasons. First, due to the lack of a condition similar to the SRP condition in VeMAC, when two vehicles having acquired a time slot enter the communication range of each other, one of them releases its time slot even if no merging collision happens. Second, as mentioned in [4], a node which needs to acquire a time slot should attempt transmission in the next available time slot with probability p_{acc}. For a certain time slot, the optimal probability $p_{opt} = 1/N_c$, where N_c is the number of contending nodes attempting to acquire this time slot [4]. However, since N_c is not known to any of the contending nodes, each contending node x sets $N_c = N_{max} - N_{succ}^x$, where N_{max} is the maximum number of nodes which can exist in a THS and N_{succ}^x is the number of nodes in the two-hop neighbourhood of node x which have successfully acquired a time slot as derived from the framing information received by node x [4]. This estimation of N_c is far from accurate. The reason is that, if a node x detects that N_{succ}^x nodes have successfully acquired a time slot, this does not mean at all that there are $N_{max} - N_{succ}^x$ nodes which need to acquire a time slot in the two-hop neighbourhood of node x. Also, even if there are exactly $N_{max} - N_{succ}^x$ contending nodes, they do not necessarily contend for the same time slots since each of the nodes may belong to a different set of THSs. Additionally, N_{max} is not constant since it depends on parameters such as the inter-vehicle distance and the number of lanes which considerably vary based on the scenario (i.e., highway, city, urban, sub-urban, or rural areas).

In terms of communication over the control channel, the main similarities and differences between the VeMAC and ADHOC MAC protocols can be summarized as follows. Both protocols are based on TDMA, work over the physical layer of different standards (such as the IEEE 802.11), and achieve an efficient multihop broadcast service [7, 8], as well as a reliable one-hop broadcast service without the hidden terminal problem. Also, they both require each node to periodically announce the time slots used by all nodes within its one-hop neighbourhood. However, as will be shown in Sect. 3.5.4, the VeMAC protocol significantly outperforms the ADHOC MAC protocol, thanks to the following three main features: the reduction of the access collision rate by using fixed time frames (versus sliding frames in ADHOC MAC) and a new method for the nodes to access the available time slots, the reduction of the merging collision rate by assigning disjoint sets of time slots to vehicles moving in opposite direction and to RSUs, and the SRP condition which prevents the nodes from unnecessarily releasing their time slots when they just enter the communication range of each other. These advantages of VeMAC come in addition to being a multichannel protocol more suitable for the DSRC spectrum as compared to the single channel ADHOC MAC protocol. On the other hand, the VeMAC protocol requires frame synchronization, which is not needed by the ADHOC MAC protocol (due to the use of sliding frames). The frame synchronization can be achieved by using the GPS 1PPS signal with an integer number of frames in each second, as discussed in Sect. 2.2 and 2.4.

Based on the two limitations of the typical ADHOC MAC protocol [4], two more versions of ADHOC MAC are considered in the simulations: the ADHOC-enhanced (AE) and the ADHOC-optimal (A-opt). The AE protocol eliminates the first limitation of ADHOC MAC by using a condition similar to the SRP condition

Table 3.2 The simulated
protocols [1]

Protocol	Abbreviation
VeMAC with $\tau = \infty$	V-inf
VeMAC with $\tau = 0$	V0
ADHOC MAC as in [4]	ADHOC
ADHOC-enhanced	AE
ADHOC-optimal	A-opt

Fig. 3.12 The number of
vehicles acquiring a time slot
for the three ADHOC MAC
versions in the highway
scenario, at THSO = 0.6 (i.e.,
60 vehicle/THS) [1]

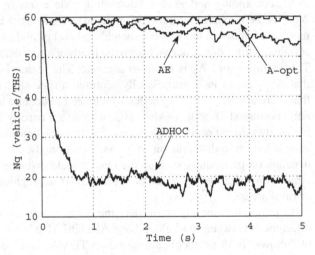

of VeMAC. More precisely, a node x does not release its time slot based on a packet
received from a node y unless node x has previously received a packet from node
y, i.e., unless node y is included in the framing information [4] constructed by node
x. For both ADHOC MAC and AE, the probability of accessing an available time
slot by a contending node x is $p_{acc} = \frac{1}{L-N_{succ}^x}$. Note that, N_{max} is replaced by L
(i.e., the maximum number of slots available for a THS) as it is not mentioned in [4]
how to determine N_{max}. To evaluate the second limitation of ADHOC MAC, the
A-opt protocol is implemented. The A-opt is similar to the AE protocol with the
difference that, for each time slot, each contending node is aware of the number of
contending nodes N_c within its two-hop neighbourhood and sets $p_{acc} = p_{opt} = \frac{1}{N_c}$.
Note that this awareness of N_c is provided by the simulator and cannot be achieved in
reality. Hence, the A-opt is not a realistic protocol, it just represents an upper bound
on the performance of ADHOC MAC. The five MAC protocols under consideration
are summarized in Table 3.2.

To demonstrate the difference among the three ADHOC MAC versions, Fig. 3.12
shows the total number of vehicles successfully acquiring a time slot, denoted by N_q,
in the first five seconds of the simulation in the highway scenario. For the ADHOC
protocol, due to the lack of the SRP condition, N_q drops from 60 to 20 vehicle/THS
in the first second of the simulation. Also, each vehicle which releases its time slot

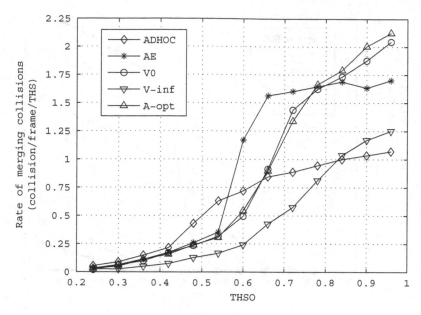

Fig. 3.13 The rate of merging collisions in highway [1]

in the first second cannot quickly acquire a new one due to the inexact probability of accessing an available time slot. For this reason, N_q remains below 20 vehicle/THS at the end of the five seconds. Unlike ADHOC MAC, in the AE protocol, the sudden decrease in N_q is eliminated thanks to the SRP condition. For this protocol, N_q decreases gradually and reaches 54 vehicle/THS at the end of the five seconds. On the other hand, the A-opt protocol does not show any decrease in N_q at the end of the five seconds since it can control the access collisions by using the optimal probability p_{opt} for accessing the available time slots. Similar behaviours of the three ADHOC MAC versions were seen in the city scenario.

3.5.4 Simulation Results

3.5.4.1 Highway Scenario

Figure 3.13 shows the rate of merging collisions for all the MAC protocols under consideration. The V-inf protocol achieves a low rate of merging collisions since it assigns disjoint sets of time slots to vehicles moving in opposite directions. The V0 and A-opt protocols have almost the same rate of merging collisions for different THSO values. Note that for a high THSO, the ADHOC protocol provides a low rate of merging collision, even less than the V-inf protocol, due to a small number of nodes which successfully acquire a time slot as compared to other protocols

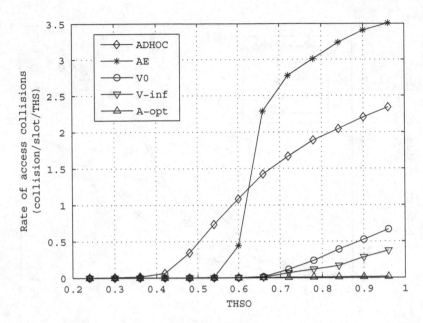

Fig. 3.14 The rate of access collisions in highway [1]

(recall that, by definition, a merging collision happens only among the nodes which are successfully acquiring a time slot).

The rate of access collisions is shown in Fig. 3.14 for all the protocols. As expected, the A-opt protocol shows a considerably smaller rate of access collisions than both ADHOC and AE protocols, which verifies the inefficiency of both protocols in determining the probability of accessing an available slot. Due to the ability of the V-inf protocol to decrease the rate of merging collisions, as shown in Fig. 3.13, it also achieves a less rate of access collisions than that of the V0 protocol. The reason is that, each merging collision generates access collisions, especially for a high THSO, until each node which released its time slot reacquires a new one. Both VeMAC protocols (V-inf and V0) provide a rate of access collisions which is slightly higher than that of the A-opt protocol but significantly lower than the rates provided by the ADHOC and AE protocols especially for a high THSO.

Figure 3.15 shows the Tx throughput for all the protocols. Because of the limitations discussed in Sect. 3.5.3, the performance of the ADHOC protocol is the lowest among all the MAC protocols for all the THSO values. The AE protocol has better performance than the ADHOC protocol, but its Tx throughput decreases for a high THSO due to its inability to handle the access collisions. For a THSO < 0.7, the V-inf and V0 protocols have almost the same Tx throughput, while for a THSO > 0.7, the V-inf protocol starts to perform better than the V0 protocol. Both protocols outperform the AE and ADHOC protocols for all the THSO values, and the Tx throughput of the V-inf is slightly less than the unrealistic A-opt protocol for a THSO > 0.7.

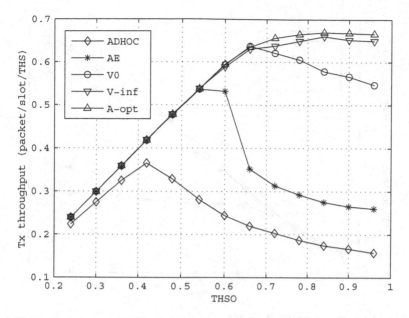

Fig. 3.15 The Tx throughput in highway [1]

The Rx throughput is shown in Fig. 3.16. It is clear that, the V-inf and V0 protocols achieve a higher Rx throughput than both of the AE and ADHOC protocols for all the THSO values. For instance, at THSO = 0.78, the V-inf protocol provides an Rx throughput of 51 packet/slot/THS as opposed to only 21 packet/slot/THS in the case of the ADHOC protocol (i.e., a 143 % increase in the Rx throughput). Note that, for a high THSO, even if the Tx throughput remains constant or slightly decreases, the Rx throughput continues increasing. The reason is that, for the same Tx throughput, when the number of vehicles on the highway segment increases (i.e., when the THSO increases), more vehicles can receive packets since all the packets transmitted are of broadcast type. Similar to the Tx throughput, the V-inf protocol provides a slightly less Rx throughput than the A-opt protocol. For the range of THSO considered in the highway, the maximum relative difference[3] between the Rx throughput of the V-inf and A-opt protocols is approximately 3.9 % (achieved at THSO = 0.72).

3.5.4.2 City Scenario

The rate of merging collision in the city scenario is shown in Fig. 3.17 for all the protocols. It is noted that, the relative difference between the rate of merging collision provided by the V-inf protocol and that provided by the V0 protocol is reduced as

---

[3] The relative difference between two values x and y is defined as $\frac{|x-y|}{\min(x,y)}$.

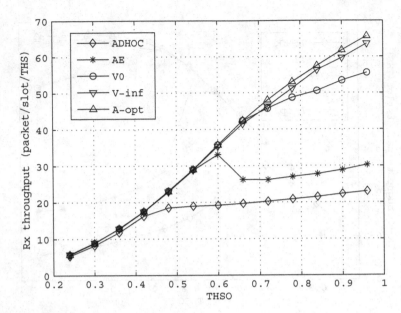

Fig. 3.16 The Rx throughput in highway [1]

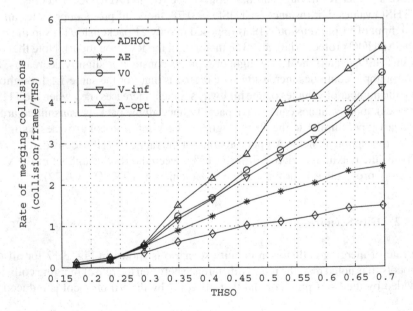

Fig. 3.17 The rate of merging collisions in city [1]

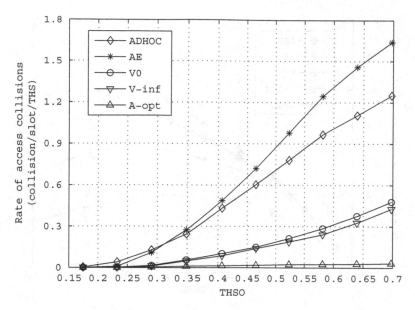

Fig. 3.18 The rate of access collisions in city [1]

compared to the highway scenario. For instance, at a THSO = 0.7 in the highway
scenario, the V0 protocol shows approximately 150 % higher rate of merging collision
than the V-inf protocol, as opposed to only an 8 % increase in the city scenario at
the same THSO. The reason is that, in the city scenario, the V-inf protocol suffers
from the merging collisions near the junction areas due to vehicles which change their
moving direction. This kind of merging collision does not exist with the V-inf protocol
when employed in the highway scenario (the merging collisions only happens among
vehicles moving in the same direction). The close rate of merging collisions of both
V-inf and V0 protocols also results in a close rate of access collisions, as shown in
Fig. 3.18. Similar to the highway scenario, both V-inf and V0 protocols provide a
rate of access collision which is higher than that of the A-opt protocol but lower than
those provided by the AE and ADHOC protocols.

The Tx throughput and Rx throughput are shown in Figs. 3.19 and 3.20 respectively. The V-inf and V0 protocols have the same performance for a THSO < 0.5,
while the V-inf protocol performs slightly better for a THSO > 0.5. Unlike the
highway scenario, where the A-opt and V-inf protocols have very close Tx and Rx
throughputs, in the city scenario the A-opt outperforms the V-inf protocol. This
outperforming is a result of the excess merging collisions that the V-inf protocol
experiences in the city scenario due to vehicles which change their moving directions. However, similar to the highway scenario, both V-inf and V0 protocols provide
higher Tx and Rx throughputs than the AE and ADHOC protocols.

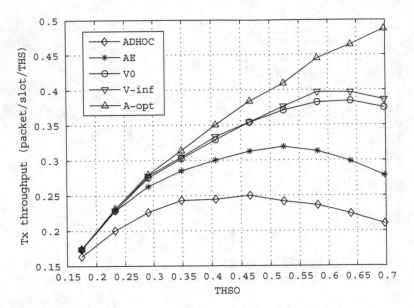

Fig. 3.19 The Tx throughput in city [1]

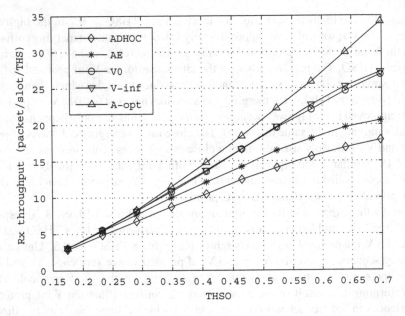

Fig. 3.20 The Rx throughput in city [1]

3.6 Summary

This chapter presents VeMAC, a novel multichannel MAC protocol based on TDMA for VANETs. How the periodic and event-driven safety messages are queued and served by the VeMAC protocol is described, and the techniques employed by the protocol for accessing the CCH and SCHs are explained. Mathematical analysis and computer simulations are presented to evaluate the performance of the VeMAC protocol in highway and city scenarios, in comparison with different versions of the ADHOC MAC protocol. Simulation results show that VeMAC provides a smaller rate of transmission collisions, which results in a significantly higher throughput on the CCH, as compared with that provided by ADHOC MAC. This outperforming of the VeMAC protocol is due to the following three main features: the reduction of the access collision rate by using fixed time frames (versus sliding frames in ADHOC MAC) and a new method for the nodes to access the available time slots, the reduction of the merging collision rate by assigning disjoint sets of time slots to vehicles moving in opposite direction and to RSUs, and the SRP condition which prevents the nodes from unnecessarily releasing their time slots when they just enter the communication range of each other. In addition, the multichannel VeMAC protocol is more suitable for the DSRC spectrum (divided into seven channels) as compared to the single channel ADHOC MAC protocol. Chapter 4 focuses on the delay performance achieved by VeMAC in delivering periodic and event-driven safety messages, and compares its performance with that of the IEEE 802.11 p standard via computer simulations in a realistic city scenario.

References

1. H. A. Omar, W. Zhuang, and L. Li, "VeMAC: A TDMA-based MAC protocol for reliable broadcast in VANETs," *IEEE Transactions on Mobile Computing*, vol. 12, no. 9, pp. 1724–1736, Sept. 2013.
2. H. A. Omar, W. Zhuang, A. Abdrabou, and L. Li, "Performance evaluation of VeMAC supporting safety applications in vehicular networks," *IEEE Transactions on Emerging Topics in Computing*, vol. 1, no. 1, pp. 69–83, Jun. 2013.
3. H. A. Omar, W. Zhuang, and L. Li, "VeMAC: a novel multichannel MAC protocol for vehicular ad hoc networks," in *Proc. IEEE International Conference on Computer Communications (INFOCOM) Workshop on Mobility Management in the Networks of the Future World (MobiWorld 2011)*, Apr. 2011, pp. 413–418.
4. F. Borgonovo, A. Capone, M. Cesana, and L. Fratta, "ADHOC MAC: New MAC architecture for ad hoc networks providing efficient and reliable point-to-point and broadcast services," *Wireless Networks*, vol. 10, pp. 359–366, Jul. 2004. [Online]. Available: http://dx.doi.org/10.1023/B:WINE.0000028540.96160.8a.
5. "Vehicle safety communications project task 3 final report," The CAMP Vehicle Safety Communications Consortium, Tech. Rep. DOT HS 809 859, Mar. 2005.
6. "A recursive solution to an occupancy problem resulting from TDM radio communication application," *Applied Mathematics and Computation*, vol. 101, no. 1, pp. 1–3, Jun. 1999.

7. F. Borgonovo, L. Campelli, M. Cesana, and L. Coletti, "MAC for ad-hoc inter-vehicle network: services and performance," in *Proc. IEEE 58th Vehicular Technology Conference (VTC 2003-Fall)*, vol. 5, Oct. 2003, pp. 2789–2793.
8. H. A. Omar, W. Zhuang, and L. Li, "On Multihop Communications For In-Vehicle Internet Access Based On a TDMA MAC Protocol," *Proc. IEEE International Conference on Computer Communications (INFOCOM)*, Apr. 2014.

Chapter 4
VeMAC Performance Evaluation for VANET Safety Applications

This chapter investigates how the VeMAC protocol can deliver both periodic and event-driven safety messages in VANETs, by presenting a detailed delivery delay analysis, including queueing and service delays, for both types of safety messages [1, 2]. The probability mass function of the service delay is first derived, then the D/G/1 and M/G/1 queueing systems are used to calculate the average queueing delay of the periodic and event-driven safety messages respectively. As well, a comparison between the VeMAC protocol and the IEEE 802.11 p standard [3] is presented via extensive simulations using the network simulator ns-2 [4] and the microscopic vehicle traffic simulator VISSIM [5]. A real city scenario is considered and different performance metrics are evaluated, including the network goodput, protocol overhead, channel utilization, protocol fairness, probability of a transmission collision, and message delivery delay. Although the VeMAC is a multichannel protocol, this chapter focuses only on the operation of the VeMAC on the CCH, over which the high priority periodic and event-driven safety messages under consideration are transmitted.

4.1 Delay Analysis

The total delay that a safety message experiences on the CCH before reaching all the one-hop neighbours consists of five components: (1) *upper layers delay* from the time that a safety message is generated at the application layer until it is assigned to one of the two queues in Fig. 3.1, including the fragmentation time of periodic safety messages; (2) *queueing delay* between the time that a safety message (or a fragment of a safety message) is assigned to one of the queues in Fig. 3.1 and the time that it becomes the head of line (HOL); (3) *access delay* from the time that a safety message (or a fragment of a safety message) becomes the (HOL) until the start of its transmission. This delay is mainly the time spent by the transmitting node waiting for one of its acquired periodic or event-driven time slots; (4) *transmission duration* of a safety packet; (5) *propagation delay* until the safety packet completely reaches the farthest one-hop neighbour. The *upper layers delay* and *propagation delay* are

© The Author(s) 2014

H. A. Omar, W. Zhuang, *Time Division Multiple Access For Vehicular Communications*,
SpringerBriefs in Computer Science, DOI 10.1007/978-3-319-09504-2_4

not considered in the following analysis since they are negligible as compared to the other delay components. The *transmission duration* of any safety packet is assumed to be equal to the duration of one time slot. Note that, the duration of one time slot represents the maximum *transmission duration* which can be experienced by a safety packet on the CCH. However, the difference between the maximum and actual *transmission durations* (fraction of a time slot) is negligible as compared to the *queueing delay* and *access delay* (multiple time slots). The sum of the *access delay* and *transmission duration* is referred to as the *service delay*. To simplify the analysis of the *service delay* and *queueing delay*, denoted by W_s and W_q respectively, we assume that a node releases its periodic or event-driven time slot(s) and acquires a new one(s) after the transmission of each periodic or event-driven safety packet respectively. This assumption guarantees that the *service delays* of the successive periodic and event-driven safety messages assigned to the two queues in Fig. 3.1 form two sequences of independent and identically distributed (i. i. d.) random variables, which is a necessary condition for the application of the D/G/1 and M/G/1 queuing systems in Sect. 4.1.2. The assumption is reasonable in scenarios with high rates of access collisions and merging collisions, where the nodes frequently release their time slots and acquire new ones. The number of periodic and event-driven time slots, k_p and k_e, that a node can access per frame are assumed to be constant. The total delay, denoted by W, is the sum of W_s and W_q, and all delays are represented in the unit of a time slot. For any discrete random variable X, the probability mass function (PMF) and the cumulative distribution function (CDF) are denoted by f_X and F_X respectively, while the first and second moments are denoted by $\overline{X}$ and $\overline{X^2}$ respectively. If random variable X takes only non-negative integer values, its probability generating function (PGF) is denoted by $G_X(z) = \overline{z^X} = \sum_x f_X(x)z^x$, while $G'_X(z)$ denotes $\frac{d}{dz}G_X(z)$. The *service delay* and *queueing delay* are considered separately in Sects . 4.1.1 and 4.1.2 in the following. The accuracy of the analysis in this section under the simplified assumptions is studied via MATLAB simulations in Sect. 4.2.2.

4.1.1 Service Delay

Since the VeMAC protocol serves the two queues in Fig. 3.1 independently using the k_p and k_e time slots, the PMF f_{W_s} is similar for both queues and differs only due to the difference between the k_p and k_e values. Hence, the PMF f_{W_s} is derived in a generic way (i.e., irrespective of the type of the transmitted safety message) given that the transmitting node is accessing k time slots per frame. For the periodic and event-driven safety messages, the PMF f_{W_s} can be calculated just by replacing k in the generic f_{W_s} with k_p and k_e respectively. Let random variable J denote the index of the time slot at the start of which a safety message becomes the HOL. Note that, since the transmission delay is equal to 1, if the inter-arrival time of periodic safety messages is an integer value, and assuming that the first message arrives at the start of a time slot, it is guaranteed that a periodic message always becomes the HOL at the *start* of a time slot. On the other hand, due to random arrivals of event-driven safety messages with non-integer inter-arrival times, it is possible that, when the queue is empty, an arriving event-driven message becomes the HOL *within* the duration of a certain time slot. In this case, we neglect a fraction of time slot in the calculation of the *service delay* and assume that the event-driven message becomes the HOL at the start of the next slot. Hence, the *service delay* W_s can take only integer values ranging from 1 to $L - k + 1$. The calculation of $f_{W_s}(i), i = 1, \ldots, L - k + 1$, is considered separately for the two extreme values of the split up parameter, $\tau = 0$ and $\tau = \infty$.

4.1.1.1 $\tau = 0$

In this case, if a safety message becomes the HOL at the start of time slot j, the transmitting node can be accessing any k of the L time slots following (and including) time slot j with equal probabilities. Hence,

$$p\left(W_s = i | J = j\right) = \frac{C_{k-1}^{L-i}}{C_k^L},$$

$$1 \leq k \leq L, 1 \leq i \leq L - k + 1, 0 \leq j \leq L - 1$$

where $C_k^n = \frac{n!}{(n-k)!k!}$. The denominator is the number of ways that the transmitting node can access k time slots among the L time slots following (and including) time slot j, while the numerator is the number of ways that one of the k time slots that the node is accessing is the ith time slot starting from j, denoted by $j_a = (j+i-1) \bmod L$, and the remaining $k - 1$ time slots are among the $L - i$ time slots following time slot j_a. In other words, the numerator is the number of ways that the node is accessing the ith time slot starting from j but not any of the $i - 1$ time slots following (and including) time slot j. Note that, with $\tau = 0$, the probability $p(W_s = i | J = j)$ is independent of the value of j since the transmitting node is allowed to access all the available time slots in a frame with equal probabilities. Hence,

$$f_{W_s}(i) = \sum_{j=0}^{L-1} p(W_s = i | J = j) \times f_J(j) = \sum_{j=0}^{L-1} \frac{C_{k-1}^{L-i}}{C_k^L} \times f_J(j) = \frac{C_{k-1}^{L-i}}{C_k^L},$$

$$1 \leq i \leq L - k + 1, 1 \leq k \leq L.$$

4.1.1.2 $\tau = \infty$

Consider that a node is moving in one of the left directions. When a safety message becomes the HOL at the start of time slot j, the transmitting node can be accessing any k time slots in set $\mathcal{L}$ with equal probabilities. There is no probability that the node accesses any of the time slots in sets $\mathcal{R}$ and $\mathcal{F}$. Hence, unlike the $\tau = 0$ case, the probability $p(W_s = i | J = j)$ depends on the value of j.

a) For $|\mathcal{L}| \leq j \leq L - 1$, we have

$$
p(W_s = i | J = j) = \begin{cases} \frac{C_{k-1}^{|\mathcal{L}|-[i-(L-j)]}}{C_k^{|\mathcal{L}|}}, & L - j + 1 \leq i \leq L - j + 1 + |\mathcal{L}| - k, \\ & 1 \leq k \leq |\mathcal{L}|, \\ 0, & \text{elsewhere.} \end{cases}
$$

The denominator represents the total number of ways that the node can access k slots among the $|\mathcal{L}|$ time slots, while the numerator represents the number of ways which result in W_s equal to i. Note that, the smallest possible value of W_s is $L - j + 1$, since $j \in \mathcal{R} \cup \mathcal{F}$ while the node cannot access any time slot in set $\mathcal{R} \cup \mathcal{F}$.

b) For $0 \leq j \leq |\mathcal{L}| - 1$, we have the following two cases

- If $j < k$, we have $W_s \leq |\mathcal{L}| - k + 1$, since at least one of the k time slots that the node is accessing is among the next $|\mathcal{L}| - j$ time slots starting from time slot j. Then

$$
p(W_s = i | J = j) = \begin{cases} \frac{C_{k-1}^{|\mathcal{L}|-i}}{C_k^{|\mathcal{L}|}}, & 1 \leq i \leq |\mathcal{L}| - k + 1, \\ 0, & \text{elsewhere.} \end{cases}
$$

- If $j \geq k$, there is a probability that the k time slots that the node is accessing are all before time slot j, which results in W_s taking values between $L - j + 1$ and $L - k + 1$. Hence

$$
p(W_s = i | J = j) = \begin{cases} \frac{C_{k-1}^{|\mathcal{L}|-i}}{C_k^{|\mathcal{L}|}}, & 1 \leq i \leq |\mathcal{L}| - j, \\ \frac{C_{k-1}^{L-i}}{C_k^{|\mathcal{L}|}}, & L - j + 1 \leq i \leq L - k + 1, \\ 0, & \text{elsewhere.} \end{cases}
$$

Given $p(W_s = i | J = j)$ for all $0 \leq j \leq L - 1$, we have

$$
f_{W_s}(i) = \sum_{j=0}^{L-1} p(W_s = i | J = j) \times f_J(j),
$$

$$
1 \leq i \leq L - k + 1, 1 \leq k \leq |\mathcal{L}|.
$$

For a node moving in a left direction, we assume that

$$
f_J(j) = \begin{cases} \frac{1}{|\mathcal{L}|}, & 0 \leq j \leq |\mathcal{L}| - 1, \\ 0, & \text{elsewhere.} \end{cases}
$$

This assumption means that, first, a safety message cannot become the HOL at the start of time slots in set $\mathcal{R} \cup \mathcal{F}$ and, second, a safety message becomes the HOL at the start of time slots in set $\mathcal{L}$ equally likely. Note that, although the transmitting node is not allowed to access time slots in set $\mathcal{R} \cup \mathcal{F}$, a safety message still can become the HOL at the start of a time slot belonging to this set, e.g., when a message arrives at the start of a time slot $j \in \mathcal{R} \cup \mathcal{F}$ and finds the queue empty. The same procedure in this subsection can be used to derive f_{W_s} for a node moving in a right direction or for an RSU.

4.1.2 Queueing Delay

Although the PMF of the *service delay* is the same for periodic and event-driven safety messages, their *queueing delays* are different due to different arrival patterns for the two different types of safety messages.

4.1.2.1 Event-driven Safety Messages

As mentioned in Chap. 1, the event-driven safety messages are triggered by certain events such as a sudden brake, road feature notification, approaching an emergency vehicle, etc. Given the variety of such events, it is reasonable to assume that their arrival process has independent and stationary increments, with no group arrivals. That is, the numbers of events occurring in disjoint time intervals are independent, the PMF of the number of events occurring in a time interval only depends on the length of the interval, and there is no simultaneous arrival of events. Based on these properties, the arrival process of the event-driven safety messages can be modeled by a Poisson process with rate parameter λ message/slot. Hence, the event-driven safety message queue in Fig. 3.1 is an M/G/1 queue with the *service delay* distribution f_{W_s} as derived in Sect. 4.1.1. Consequently, provided that $\overline{W_s} < \frac{1}{\lambda}$, which is the necessary and sufficient condition for stability of the event-driven safety message queue [6], by applying the P-K formula [7], we have

$$\overline{W_q} = \frac{\lambda \overline{W_s^2}}{2(1 - \lambda \overline{W_s})}.$$

4.1.2.2 Periodic Safety Messages

Based on the assumption of fixed-size periodic safety messages (Sect. 2.1), the number of fragments of a periodic safety message is assumed to be fixed for a given node. If n_f denotes the number of fragments of a periodic safety message for a certain node, the arrival of each periodic safety message results in a simultaneous arrival of n_f fragments in the periodic safety message queue in Fig. 3.1. Consequently, this queue can be modeled as a D/G/1 queue with fixed-size batch arrivals. Hence, the *queueing delay* that a tagged fragment of a periodic safety message experiences consists of two components: the delay since the batch (to which the tagged fragment belongs) enters the queue until the first fragment of the batch becomes the HOL, plus the *service delay* of all the fragments queued before the tagged fragment within the batch. The two components of the *queueing delay* are independent and denoted by W_{q_1} and W_{q_2} respectively. Let integer I denote the inter-arrival time of periodic safety messages, i.e., the batch inter-arrival time. The PGF of the *service delay* of one batch, denoted by W_b, is

$$G_{W_b}(z) = (G_{W_s}(z))^{n_f}.$$

Hence, provided that $\overline{W_b} = G'_{W_b}(1) < I$, which is the necessary and sufficient condition for stability of the periodic safety message queue [6], the PGF of W_{q_1} can be calculated as follows [8, 9]

$$G_{W_{q_1}}(z) = \frac{\xi\left[\prod_{i=1}^{I-1}(z - z_i)\right](z - 1)}{z^I - G_{W_b}(z)}$$

where

$$\xi = \lim_{z \to 1} \frac{z^I - G_{W_b}(z)}{\left[\prod_{i=1}^{I-1}(z - z_i)\right](z - 1)}$$

and complex numbers $z_1, z_2, \ldots, z_{I-1}$ are the roots of the function $z^I - G_{W_b}(z)$, which are on or inside the unit circle but not equal to 1. The PGF, $G_{W_{q_2}}(z)$, can be calculated by noting that $W_{q_2} = \sum_{i=0}^{N_f} W_s$, where N_f is a random variable representing the number of fragments queued before the tagged fragment within the batch. Since the tagged fragment can be any fragment within the batch with equal probabilities, $f_{N_f}(i) = \frac{1}{n_f}, i = 0, \ldots, n_f - 1$, and $G_{N_f}(z) = \frac{1}{n_f}\sum_{i=0}^{n_f-1} z^i$. Hence, by using the law of total expectation,

$$G_{W_{q_2}}(z) = G_{N_f}(G_{W_s}(z)).$$

Consequently,

$$G_{W_q}(z) = G_{W_{q_1}}(z) \times G_{W_{q_2}}(z)$$

$$\overline{W_q} = G'_{W_q}(1).$$

4.2 Numerical Results

4.2.1 Analytical Results

We use MATLAB R2011b and the Symbolic Math Toolbox V5.7 for the calculation of the average delays as described in Sect. 4.1. Figures 4.1a and b show F_{W_s} for a node moving in a left direction with $\tau = 0$ and $\tau = \infty$ respectively. The main difference

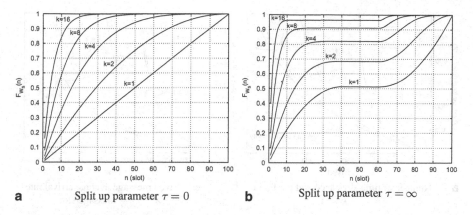

Fig. 4.1 The CDF of the *service delay*, F_{W_s}, for a node moving in a left direction with 100 time slots per frame and 40 time slots associated with the left direction, i.e., $L = 100$ and $|\mathcal{L}| = 40$ [1]. **a** Split up parameter $\tau = 0$. **b** Split up parameter $\tau = \infty$

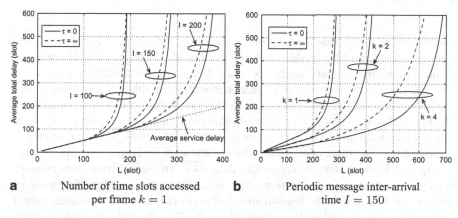

Fig. 4.2 The average total delay, $\overline{W}$, of a single-fragment periodic message ($n_f = 1$) for a node moving in a left direction with 40 % of the time slots associated with the left direction, i.e., $|\mathcal{L}| = 0.4\,L$ [1]. **a** Number of time slots accessed per frame $k = 1$. **b** Periodic message inter-arrival time $I = 150$

between the two cases is that, when $\tau = \infty$, $F_{W_s}(n)$ remains constant for a certain range of n. With $\tau = \infty$, the node can only access time slots in set $\mathcal{L}$. As a result, there should be a range of n where $f_{W_s}(n) = 0$. For instance, if $k = 2$, $L = 100$, and $|\mathcal{L}| = 40$, $f_{W_s}(n) = 0, \forall n \in \{40, \dots, 61\}$.

Figure 4.2a shows the average total delay $\overline{W}$ of a periodic safety message with $n_f = 1$ (a typical case for vehicles) for a node moving in a left direction with $k = 1$. Both $\tau = 0$ and $\tau = \infty$ cases are plotted in Fig. 4.2a for various I values. Although the $\tau = 0$ and $\tau = \infty$ cases have different F_{W_s} (in Figs. 4.1a and b), when $k = 1$, both τ values result in the same $\overline{W_s}$, which is represented by the straight line in Fig. 4.2a. As shown in Fig. 4.2a, if $L \leq I$, $\overline{W}$ is the same as $\overline{W_s}$ since each safety

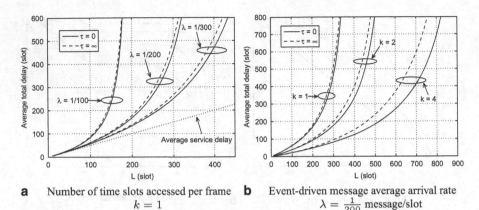

a   Number of time slots accessed per frame   **b**   Event-driven message average arrival rate
$$k = 1$$ $$\lambda = \tfrac{1}{200} \text{ message/slot}$$

Fig. 4.3 The average total delay, $\overline{W}$, of an event-driven safety message for a node moving in a left direction with 40 % of the time slots associated with the left direction, i.e., $|\mathcal{L}| = 0.4\,L$ [1]. **a** Number of time slots accessed per frame $k = 1$. **b** Event-driven message average arrival rate $\lambda = \tfrac{1}{200}$ message/slot

message is served before the next one arrives, i.e., $\overline{W_q} = 0$. When $L > I$, the queueing component W_q is added to the total delay W, and the value of $\overline{W}$ continues to increase with L and approaches ∞ when L tends to the instability value L^* at $\overline{W_s} = I$. Eventually, the value of L^* increases with the number of time slots, k, that the node is allowed to access per frame. To illustrate the effect of k on the total delay W, Fig. 4.2b shows $\overline{W}$ for $I = 150$ and different k values. As shown in Fig. 4.2b, while a frame duration $L = 300$ results in instability for the $k = 1$ case, when k is increased to 2, the value of $\overline{W}$ remains below 200 slots for both $\tau = 0$ and $\tau = \infty$.

Figure 4.3a illustrates the average total delay $\overline{W}$ of an event-driven safety message for a node moving in a left direction with $k = 1$. Unlike the periodic safety message case in Fig. 4.2a, due to the Poisson arrival of event-driven safety messages, even if $L \leq \tfrac{1}{\lambda}$, the queueing delay $W_q > 0$ and $\overline{W} > \overline{W_s}$. The effect of k on the total delay of event-driven safety messages is shown in Fig. 4.3b for $\lambda = \tfrac{1}{200}$ message/slot.

4.2.2 Simulation Results

Computer simulations have been conducted using MATLAB to simulate the two queues in Fig. 3.1. The objectives of the simulations are to study the impact of the assumption on f_J for the $\tau = \infty$ case, the influence of neglecting a fraction of time slot in the derivation of f_{W_s} for the event-driven safety messages, and the effect of the numerical errors such as in calculating the roots of $z^I - G_{W_b}(z)$. Fig.4.4a shows $\overline{W_s}$ and $\overline{W_q}$ of a periodic safety message with $n_f = 1$ for a node moving in a left direction with $k = 1, \tau = 0$, and $|\mathcal{L}| = 0.4\,L$. The same parameter values are used in Fig. 4.4b to illustrate the average queueing delay $\overline{W_q}$ of an event-driven safety

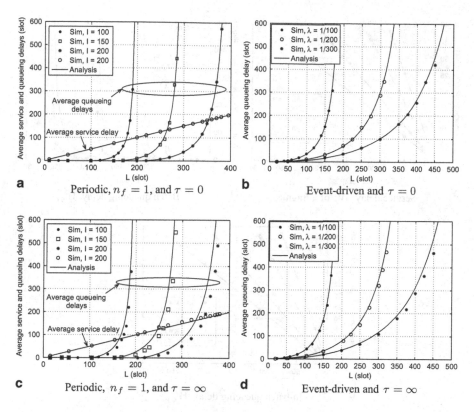

Fig. 4.4 Analysis and simulation (*Sim*) results of the average delays for a node moving in a left direction with $k = 1$ and $|\mathcal{L}| = 0.4\,L$ [2]. **a** Periodic, $n_f = 1$, *and* $= 0$. **b** Event-driven and $\tau = 0$. **c** Periodic, $n_f = 1$, *and* $\tau = \infty$. **d** Event-driven and $\tau = \infty$

message. Note that, the average service delay $\overline{W_s}$ is not shown in Fig. 4.4b since it is the same as in Fig. 4.4a (W_s is independent of the arrival pattern). As shown in Figs. 4.4a and b, there is a close match between the analysis and simulation results of the delays of both periodic and event-driven safety messages. The same delays are shown in Figs. 4.4c and d for $\tau = \infty$. Unlike the $\tau = 0$ case, a slight mismatch appears between the analysis and simulation results in Figs. 4.4c and d, mainly for large I and small λ values, due to the assumption on f_J. The effect of this assumption is worse on the periodic safety messages than on the event-driven safety messages. However, the analysis and simulation results for both types of safety messages are still close to each other in Figs. 4.4c and d.

To consider a case of large-size periodic safety messages (typically for RSUs), the three delay components $\overline{W_s}$, $\overline{W_{q_1}}$, and $\overline{W_{q_2}}$ of a multi-fragment periodic safety message with $n_f = 4$ are shown respectively in Figs. 4.5a, b, and c for an RSU with $\tau = 0$ and different k values. As shown in Figs. 4.5a and c, there is a close match between the analysis and simulation results of $\overline{W_s}$ and $\overline{W_{q_2}}$. However, some

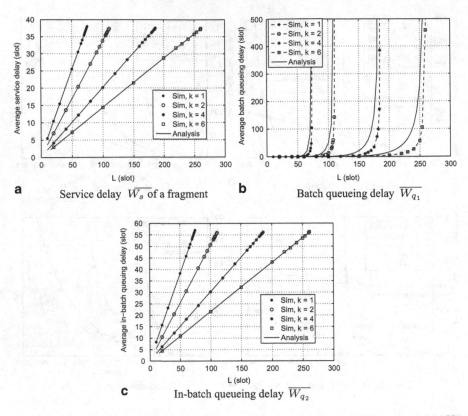

Fig. 4.5 Average delays of a periodic safety message with $n_f = 4$ for an RSU when $I = 150$, $|\mathcal{F}| = 0.4L$, and $\tau = 0$ [2]. **a** Service delay $\overline{W_s}$ of a fragment. **b** Batch queueing delay $\overline{W_{q_1}}$. **c** In-batch queueing delay $\overline{W_{q_2}}$

mismatch is noticed for $\overline{W_{q_1}}$ in Fig. 4.5b. This mismatch is the effect of numerical errors, mainly in the calculation of the roots of $z^I - G_{W_b}(z)$. The numerical errors are more significant for large n_f and L due to an increase in the degree of the polynomial $z^I - G_{W_b}(z)$, since $G_{W_b}(z) = (G_{W_s}(z))^{n_f}$ and $G_{W_s}(z)$ itself is a polynomial of degree $L - k + 1$.

4.2.3 Discussion

Based on the numerical results in Sects. 4.2.1 and 4.2.2, it is observed that the delay performance of the VeMAC with $\tau = 0$ is better than $\tau = \infty$ for both periodic and event-driven safety messages, especially for large k and I, and small λ values. If the size of the periodic safety messages broadcast by vehicles is 150 bytes, a VeMAC MTU of 675 bytes is suitable to include one periodic safety message and all the VeMAC control information which should be transmitted on the CCH. For a transmission rate of 18 Mbps, which is one of the rates supported by the IEEE 802.11 p orthogonal frequency division multiplexing (OFDM) physical layer for the 5 GHz band, the VeMAC MTU transmission time is 0.3 ms. By including guard periods and considering the physical layer overhead, a slot duration of 0.35 ms can be assumed. Given this slot duration, for the periodic safety messages of vehicles, if $I = 200$slots $= 70$ms, and each vehicle is allowed to access one periodic time slot per frame, then from Fig. 4.2a, a frame duration $L = 300$ results in an average total delay around 185 slots (65 ms) for the $\tau = 0$ case. Similarly, for the event-driven safety messages in Fig. 4.3a, if $\lambda = \frac{1}{300}$ message/slot $= 9.5$ message/s, and if the transmitting node is allowed to access only one event-driven time slot per frame, a frame duration of 300 slots results in an average delay around 250 slots (88 ms). Note that, the frame duration L represents the maximum number of time slots available for any THS in the network. For instance, if $L = 300$ slots and the transmission range is 200 m (corresponding to the maximum length of 400 m occupied by a THS on a road segment), the total number of time slots available for all the nodes on a road segment of any 400 m is equal to 300 slots. The results in this section help to determine the VeMAC parameters, such as $\tau, k_p, k_e,$ and L, used for the comparison with the IEEE 802.11 p standard as follows.

4.3 Comparison of VeMAC with IEEE 802.11 p

Computer simulations are conducted using the network simulator ns-2 [4] to evaluate the performance of the VeMAC protocol in comparison with the IEEE 802.11 p standard in broadcasting the safety messages. Periodic safety messages are generated continuously, while event-driven safety messages are generated according to an exponential ON/OFF model (i.e., the ON and OFF periods are exponentially distributed) at each node in the simulations. For the VeMAC protocol, the periodic and event-driven safety messages are queued and served as specified in Sects. 3.1 and 3.2 [1]. On the other hand, for the IEEE 802.11 p, we have employed the EDCA scheme, which assigns any MSDU to one of four different access categories (ACs) [10]. The event-driven and periodic safety messages are respectively assigned to the highest and second-highest priority ACs, i.e., AC_VO and AC_VI [3]. Two simulation scenarios are considered: a square network and a realistic city scenario [2]. For both scenarios, the ns-2 parameters are summarized in Table 4.1. The IEEE 802.11 p parameter values in Table 4.1 are as specified by the IEEE 802.11 p OFDM physical layer for the 5 GHz band [3, 10]. The carrier frequency of 5.89 GHz represents the

---

[1] A website [11] is created in order to upload the ns-2 implementation of the VeMAC protocol, including the periodic and event-driven message queues, for interested researchers.

[2] To the best of our knowledge, currently there is no benchmark vehicle mobility scenarios which can be used for the evaluation of VANET networking protocols.

Table 4.1 ns-2 simulation parameters [1]

Periodic messages		Event-driven messages				Physical layer			
Parameter	*Value*	*Parameter*	*Value*	*Parameter*	*Value*	*Parameter*	*Value*	*Parameter*	*Value*
Size	150 bytes	Size	450 bytes	Average OFF time	2 s	RXThresh	1.45683×10^{-09} w	CSThresh	8.19468×10^{-10} w
Arrival rate	10 message/s	Average ON time	1 s	Arrival rate during ON time	10 message/s	Carrier frequency	5.89 GHz	Transmission power	33 dBm
						CPThresh	10	Transmission rate	12-18 Mbps
						Antenna	Omni-directional	Channel model	free space

Higher layer protocols		VeMAC				IEEE 802.11p			
Layer	*Protocol*	*Parameter*	*Value*	*Parameter*	*Value*	*Parameter*	*Value*	*Parameter*	*Value*
Transport layer	UDP	L	275 slots	Slot duration	0.35 ms	aSlotTime	13 μs	SIFS	32 μs
Network layer	dumb agent	k_p	1	k_e	1	AC_VO CW size	3	AC_VO AIFS	58 μs
		τ	0	MTU	450-675 bytes	AC_VI CW size	7	AC_VI AIFS	71 μs
		#bits of a node ID	9	#bits of a slot index	9	Preamble length	32 μs	PLCP header length	8 μs
Simulation time: 1 min. for square network and 5 min. for city						FCS	4 bytes	Header length	32 bytes

center frequency of the DSRC channel 178 (the CCH), and the transmission power of 33 dBm is the maximum power allowed on this channel for private OBUs and RSUs as in the ASTM E2213 standard [12]. Given these values of the carrier frequency and the transmission power, the receiving threshold (RxThresh) and the carrier sensing threshold (CSThresh) in Table 4.1 result in a communication range of 150 m and a carrier sensing range of 200 m for free space propagation. The capture threshold (CPThresh) is the minimum ratio between the powers of two received signals required for the receiver to capture the signal with the higher power and discard the one with the lower power. The dumb agent used in the network layer just passes the data from the transport layer to the MAC layer while sending, and vice versa while receiving (since all the safety messages under consideration are single-hop broadcast messages).

In addition to the total delay (as defined in Sect. 4.1), the following performance metrics are considered:

a) goodput: the average rate of safety messages which are successfully delivered to all the one-hop neighbours;
b) channel utilization: the percentage of time that the channel is used for successful transmission of payload data (a transmission is considered successful only if it is correctly received by all the one-hop neighbours);
c) overhead: the percentage of control information relative to the total information transmitted on the channel;
d) probability of a transmission collision: the probability that a transmitted safety message experiences a collision at one or more one-hop neighbours; and
e) fairness indicator: for each node x, a metric denoted by r_x is first calculated, which represents the ratio of the number of safety messages *transmitted* by node x to the total number of safety messages *transmitted* by all nodes. The fairness indicator is the deviation (in percentage) of r_x from a fair share, s_x, that equals

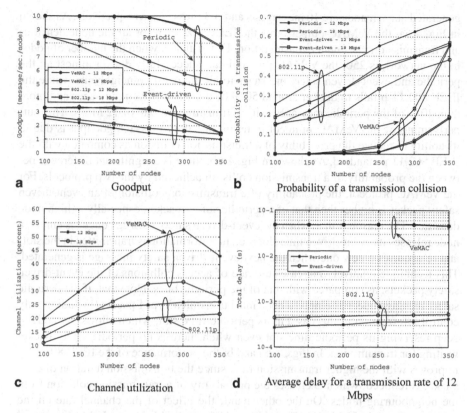

Fig. 4.6 Simulation results for the square network [1]. **a** Goodput. **b** Probability of a transmission collision. **c** Channel utilization. **d** Average delay for a transmission rate of 12 Mbps

the total number of safety messages *generated* at node x normalized by the total number of safety messages *generated* at all nodes. That is, the fairness indicator for a node x is equal to $\mid \frac{r_x - s_x}{s_x} \mid \times 100$.

All the performance metrics, except the overhead and the channel utilization, are calculated separately for the periodic and event-driven safety messages.

4.3.1 Square Network

The first scenario under consideration is a set of stationary nodes uniformly distributed in a square network with side length of 500 m. Figure 4.6a shows the periodic and event-driven message goodputs of the VeMAC and the IEEE 802.11 p protocols using two different physical layer transmission rates. Note that, based on the parameters in Table 4.1, the average rates of periodic and event-driven safety messages generated at each node are 10 messages/s and 3.3 messages/s respectively. As shown in Fig. 4.6a, the VeMAC outperforms the IEEE 802.11 p for all the node densities and transmission rates under consideration. For instance, when the number of nodes in the network is 250, the VeMAC protocol can successfully deliver almost all the periodic and event-driven safety messages to all the one-hop neighbours, while the IEEE 802.11 p fails to deliver around 50 % of the event-driven messages and more than 40 % of the periodic messages using a transmission rate of 12 Mbps. This out-performing of the VeMAC protocol in terms of safety message goodput is due to its ability to reduce the probability of a transmission collision as compared with the IEEE 802.11 p standard. As shown in Fig. 4.6b, there is a significant difference be-tween the probability of a transmission collision achieved by the two protocols. For the VeMAC protocol, the probability of a transmission collision of an event-driven safety packet is higher than that of a periodic safety packet, especially at high node densities. The reason is that, when the event-driven safety message queue is empty, a node releases its event-driven time slot (i.e., no information is transmitted in the slot) and re-acquires a new one when the next event-driven safety message is generated. This technique relatively increases the rate of access collisions of the event-driven safety packets, as compared with that of the periodic ones. Note that, if the periodic safety message queue is empty, a node must transmit a Type1 packet (including only control information in this case) in its periodic time slot, which allows the node to keep reserving its periodic time slot even when there is no periodic safety packet waiting for transmission. In Figs. 4.6a and b, the performance of the IEEE 802.11 p improves with the higher transmission rate, since the transmission duration of each packet is reduced, which decreases the probability of a transmission collision from the neighbouring nodes. On the other hand, the effect of the channel rate on the performance of the VeMAC in Figs. 4.6a and b is negligible. As the VeMAC protocol achieves a higher message goodput than the IEEE 802.11 p, it also provides a better channel utilization, as illustrated in Fig. 4.6c. The channel utilization in Fig. 4.6c improves with the lower transmission rate, due to an increase in the packet transmis-sion duration, which consequently increases the percentage of time that the channel is used for successful transmissions. When the transmission rate decreases from 18 Mbps to 12 Mbps, the channel utilization of the VeMAC protocol increases by a factor of 1.5 (the same ratio between the two transmission rates), while that of the IEEE 802.11 p increases by a factor less than 1.5, as the probability of a transmission collision also increases with the lower transmission rate.

Figure 4.6d shows the total delay of the VeMAC and the IEEE 802.11 p protocols. For both periodic and event-driven safety messages, the total delay of the VeMAC protocol is dominated by the *access delay* component, which is around 48 ms (one half the duration of a frame). At the lowest node density in Fig. 4.6d, the total delay of the periodic safety messages for the IEEE 802.11 p protocol is around 280 μs, which is the sum of the durations of one AC_VI arbitrary interframe space (AIFS) (71 μs), one periodic safety packet *transmission duration* (164 μs), and the average backoff time $\left(\frac{CW\,size}{2} \times aSlotTime = 45.5\mu s\right)$. This delay increases with the node density, due to an increase in the number of backoff cycles that a periodic safety packet encounters. The delay of the event-driven safety messages for the IEEE 802.11 p

Fig. 4.7 A snap shot of the simulations showing the road network with the simulated roads in blue, a 2D view of the intersection of University/Seagram streets, and a 3D view of the intersection of University/Westmount streets

protocol is higher than that of the periodic safety messages, due to a large size of the event-driven messages, which results in a higher *transmission duration*. Although the VeMAC has a higher total delay than the IEEE 802.11 p protocol, it is well below the 100 ms delay bound required for most of the safety applications [13].

4.3.2 City Scenario

We consider the city scenario as shown in Fig. 4.7, which consists of a set of roads around the UW campus. To simulate vehicle traffic, the microscopic vehicle traffic simulator VISSIM is employed [5, 14]. The simulator generates a vehicle trace file, which is transformed to an ns-2 scenario file using a MATLAB parser[3]. At the start of the simulation, vehicles enter the road network from every possible entry according to a Poisson process with rate λ_v. After a certain time duration t_{in}, the vehicle input to the road network is stopped, and after an additional warm up period t_w (to reduce transient state effects), the position and speed of each vehicle are recorded at the end of every simulation step. Two types of vehicles are considered: cars and buses. The two vehicle types differ mainly in the vehicle dimensions, as well as the maximum/desired acceleration and deceleration as functions of the vehicle speed. All cars and buses have the same desired speed distribution, which differs from one road to another, and during the left and right turns at intersections. Every intersection in the road network is controlled either by a traffic light, or a stop sign, based on how the intersection is controlled in reality. At signalized intersections, left turns are controlled by the traffic light controller, and right turns are allowed during the red signal phase. Before a vehicle enters an intersection area, it decides whether to turn left, turn right, or not to make any turn, according to a certain probability mass function, which differs from one intersection to another.

The car following model used is the Wiedemann 74 model [17] developed for urban traffic. A vehicle can be in one of four modes: free driving, approaching,

---

[3] Videos of the VISSIM and ns-2 simulations have been recorded and uploaded to [15] and [16] respectively.

Table 4.2 VISSIM simulation parameters [1]

Vehicle input		Desired speed (Km/h) distribution		Car following (Wiedemann 74)		Lane changing			Vehicle characteristics		
Parameter	*Value*	*Location*	*Distribution*	*Parameter*	*Value*	*Parameter*	*Lane changer*	*Trailing vehicle*	*Parameter*	*Car*	*Bus*
λ_v	1000 vehicles/hour	Ring road	$U(32, 48)$	AX	2 m	Maximum deceleration	-4 m/s^2	-3 m/s^2	Average length	4.44 m	11.54 m
t_w	5 min.	All other roads	$U(55, 72)$	BX$_{add}$	2 m	-1m/s^2 per distance	100 m	100 m	Width	1.5 m	2.5 m
t_{in}	2, 4, 6, and 7 min.	Right turns	$U(12, 18)$	BX$_{mult}$	3 m	Accepted deceleration	-1 m/s^2	-1 m/s^2	Percentage of the total # vehicles	95%	5%
# vehicles	292, 603, 839, and 948	Left turns	$U(20, 30)$	Simulation time: 5 min.		Default VISSIM maximum/desired acceleration and deceleration functions for cars and buses have been used.					

following, and braking. In each mode, the vehicle acceleration is a function of the vehicle speed, the characteristics of the driver and the vehicle, as well as the distance and the speed difference between the subject vehicle and the vehicle in front [17]. The last two variables also determine the thresholds between the four driving modes of a vehicle. The Wiedemann 74 model uses three parameters: the average standstill distance (AX), the additive part of the safety distance (BX$_{add}$), and the multiplicative part of the safety distance (BX$_{mult}$). The AX parameter is the average desired distance between stationary vehicles, and is used with the BX$_{add}$ and BX$_{mult}$ parameters to determine the desired following distance of a vehicle [17]. A vehicle can perform a lane change, either to turn left or right, or because it has a higher speed than the vehicle in front and there is more space in an adjacent lane. The lane change decision depends on the desired safety distance parameters (i.e., BX$_{add}$ and BX$_{mult}$), as well as on the speeds and decelerations of the vehicle making the lane change and the vehicle coming from behind in the destination lane. The VISSIM simulation parameters are summarized in Table 4.2.

As shown in Figs. 4.8a and b, for all the vehicle densities under consideration, the VeMAC protocol can successfully deliver almost all the periodic and event-driven safety messages to all the vehicles in the one-hop neighbourhoud. At the highest vehicle density, the VeMAC protocol achieves around 23 and 32 % higher goodput respectively in the periodic and event-driven safety message goodputs, as compared to the IEEE 802.11 p. Figure 4.8c shows the significant difference in the probability of a transmission collision achieved by the two protocols. For instance, when the number of vehicles is 839, the probability of a collision of a periodic (event-driven) safety message for the IEEE 802.11 p is around 2 order of magnitude (1.5 order of magnitude) greater than for the VeMAC protocol. One main reason of the high probability of a transmission collision for the IEEE 802.11 p is the hidden terminal problem, since for broadcast packets, no handshaking [request-to-send (RTS)/clear-to-send (CTS)] information exchange is used and no acknowledgement is transmitted from any recipient of the packet [10]. Another reason is that, although the small contention window (CW) size assigned to the AC_VO and AC_VI allows the safety packets to be transmitted with small delays, it increases the probability of a transmission collision when multiple vehicles within the same THS are simultaneously trying to broadcast

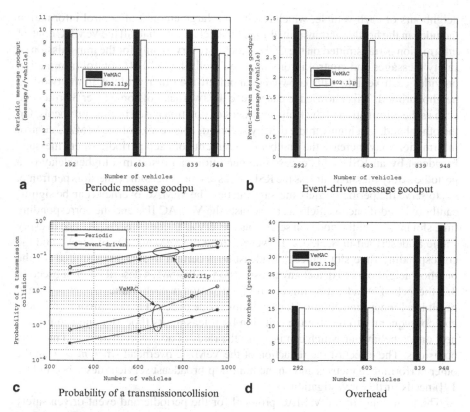

Fig. 4.8 Simulation results for the city scenario: goodput, probability of a transmission collision, and protocol overhead [1]. **a** Periodic message goodput. **b** Event-driven message goodput. **c** Probability of a transmission collision. **d** Overhead

their safety packets. Further, if a transmission collision of a broadcast packet happens, the CW size is not doubled (such as in the unicast case), as there is no collision detection without CTS and acknowledgment packets.

The reduction in the probability of a transmission collision by the VeMAC protocol, which results in the high periodic and event-driven message goodputs in Figs. 4.8a and b, is achieved at the expense of an increase in the protocol overhead as shown in Fig. 4.8d. The main source of the VeMAC overhead is that every Type1 packet transmitted by a certain vehicle x includes the set of VeMAC IDs (as indicated in Table 4.1), and the corresponding time slot indices, of each one-hop neighbour in $\mathcal{N}_x$. On the other hand, the overhead of the IEEE 802.11 p protocol is due to control information such as the frame check sequence (FCS) and the physical layer convergence procedure (PLCP) header. At low vehicle density, the overheads of the VeMAC protocol and IEEE 802.11 p are similar, as shown in Fig. 4.8d. However, when the vehicle density increases, the overhead of the IEEE 802.11 p remains the same, while that of the VeMAC protocol increases due to a large number of one-hop neighbours of each vehicle, which results in a large amount of control information included in the header of transmitted Type 1 packets. Note that, all the VeMAC control information is transmitted on the CCH, which is reserved only for the transmission of safety messages and control information. As well, the VeMAC control information provides each vehicle with knowledge about all the other vehicles in the two-hop neighbourhood. This knowledge can reduce the overhead of some layer 3 protocols, such as the elimination of the Hello messages of position based routing protocols. On the other hand, in a high vehicle density scenario, a large size of the VeMAC control information may increase the number of fragments of each periodic safety message broadcast by an RSU. This excess fragmentation can result in a higher delay of a periodic safety message, unless the RSU accesses more periodic time slots per frame, k_p, to serve the periodic safety message queue. The VeMAC overhead can be significantly reduced if each vehicle, x, broadcasts the VeMAC IDs and the corresponding time slot indices of the nodes in set $\mathcal{N}_x$ once every m frames, instead of once in every frame as described in Sect. 3.2. However, since the VeMAC IDs of set $\mathcal{N}_x$ and the corresponding time slot indices broadcast by a certain node x are required for the one-hop neighbours to detect any transmission collision, as described in Subsection 3.2, the lack of broadcasting this control information in each frame (i.e., $m > 1$) may result in a longer time duration for a colliding node to detect a transmission collision, and consequently to resolve the collision by releasing its time slot and acquiring a new one, a behaviour which can increase the rates of access collisions and merging collisions. The effect of the reduction of the VeMAC overhead when $m > 1$ on the other performance metrics and on the multihop broadcast service (described in [18, 19]) needs further investigation.

The total delay of the VeMAC protocol for the periodic and event-driven safety messages is shown in Fig. 4.9a. For both types of safety messages, the VeMAC achieves a total delay that is well below 100 ms. One reason of the relative increase in the VeMAC delays at the highest vehicle density is the high contention on the time slots among different vehicles, which may force a vehicle to delay the transmission of a safety packet until a time slot is available. To study the fairness of the VeMAC protocol, Figs. 4.9b and c show the fairness indicators of the periodic and event-driven messages respectively at the highest vehicle density under consideration. The periodic (event-driven) message fairness indicator is below 0.3 % (0.2 %) for most of the vehicles, with a maximum value of 8.3 % (6.2 %). These results indicate that, even in a high vehicle density, the VeMAC protocol allows all the vehicles to transmit their safety messages in a fair way.

4.4 Summary

This chapter presents a detailed delivery delay analysis for VANET safety messages broadcast on the CCH, based on the VeMAC protocol described in Chap. 3. Both queueing and service delays of periodic and event-driven safety messages are analyzed, by taking into consideration the size and the arrival pattern of each type of

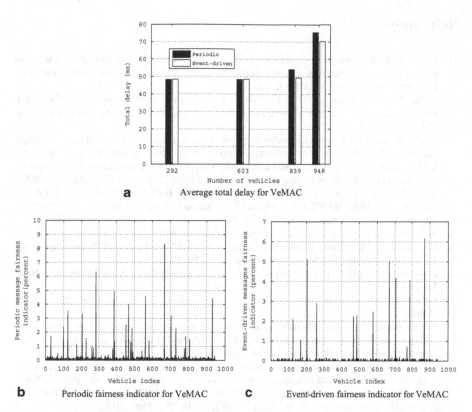

Fig. 4.9 Simulation results for the city scenario: total delay and fairness indicators of the VeMAC protocol [1]. **a** Average total delay for VeMAC. **b** Periodic fairness indicator for VeMAC. **c** Event-driven fairness indicator for VeMAC

safety messages. The delay analysis helps to determine the values of VeMAC parameters, such as $\tau, k_p, k_e,$ and L, to satisfy the delay requirement of periodic and event-driven safety applications. These protocol parameter values are used to compare the performance of VeMAC with that of the IEEE 802.11 p standard via computer simulations in a square network and in a city scenario consisting of roads around the UW campus. Simulation results show that, the VeMAC protocol has a low probability of a transmission collision, which results in a higher safety message goodput and better channel utilization, as compared with the IEEE 802.11 p standard. Also, for both types of safety messages, the VeMAC protocol achieves a total delivery delay that is well below 100 ms, which represents the maximum delay required for most of the safety applications. Additionally, by using suitable values of the VeMAC parameters, the protocol allows all the vehicles to transmit their safety messages in a fair way, even in a high vehicle density scenario. In [19], some VeMAC features, such as the knowledge of all the nodes which exist in a two-hop neighbourhood, are exploited for the design of an efficient network layer protocol.

References

1. H. A. Omar, W. Zhuang, A. Abdrabou, and L. Li, "Performance evaluation of VeMAC supporting safety applications in vehicular networks," *IEEE Transactions on Emerging Topics in Computing*, vol. 1, no. 1, pp. 69–83, Jun. 2013.
2. H. A. Omar, W. Zhuang, and L. Li, "Delay analysis of VeMAC supporting periodic and event-driven safety messages in VANETs," in *Proc. IEEE Global Communications Conference (GLOBECOM)*, Dec. 2013.
3. "IEEE standard for information technology–telecommunications and information exchange between systems–local and metropolitan area networks–specific requirements Part 11: Wireless LAN medium access control (MAC) and physical layer (PHY) specifications Amendment 6: Wireless access in vehicular environments," *IEEE Std 802.11p-2010 (Amendment to IEEE Std 802.11-2007 as amended by IEEE Std 802.11k-2008, IEEE Std 802.11r-2008, IEEE Std 802.11y-2008, IEEE Std 802.11n-2009, and IEEE Std 802.11w-2009)*, pp. 1–51, Jul. 15, 2010.
4. http://nsnam.isi.edu/nsnam/index.php/Main_Page.
5. http://vision-traffic.ptvgroup.com/en-uk/products/ptv-vissim/.
6. D. V. Lindley, "The theory of queues with a single server," *Mathematical Proceedings of the Cambridge Philosophical Society*, vol. 48, pp. 277–289, 1952.
7. D. Bertsekas and R. Gallager, *Data networks*. Upper Saddle River, NJ, USA: Prentice-Hall, Inc., 1987.
8. L. D. Servi, "D/G/1 queues with vacations," *Operations Research*, vol. 34, no. 4, pp. 619–629, 1986.
9. W. Song and W. Zhuang, "Performance analysis of probabilistic multipath transmission of video streaming traffic over multi-radio wireless devices," *IEEE Transactions on Wireless Communications*, vol. 11, no. 4, pp. 1554–1564, Apr. 2012.
10. "IEEE standard for information technology-telecommunications and information exchange between systems-local and metropolitan area networks-specific requirements - Part 11: Wireless LAN medium access control (MAC) and physical layer (PHY) specifications," *IEEE Std 802.11-2012 (Revision of IEEE Std 802.11-2007)*, pp. 1–2793, Mar. 2012.
11. http://code.google.com/p/vemac/.
12. "Specification for Telecommunications and Information Exchange Between Roadside and Vehicle Systems 5 GHz B and Dedicated Short Range Communications (DSRC) Medium Access Control (MAC) and Physical Layer (PHY) Specifications," *ASTM Standard E2213*, 2003 (2010).
13. "Vehicle safety communications project task 3 final report," The CAMP Vehicle Safety Communications Consortium, Tech. Rep. DOT HS 809 859, Mar. 2005.
14. PTV Planung Transport Verkehr AG, *VISSIM 5.40-User Manual*, 2012.
15. http://youtu.be/48daRU6ZpjI.
16. http://youtu.be/GjYWe7eLJ3s.
17. R. Wiedemann, "Modeling of RTI-Elements on multi-lane roads," in *Advanced Telematics in Road Transport, Proc. the Drive Conference*, Feb. 1991.
18. H. A. Omar, W. Zhuang, and L. Li "VeMAC: A TDMA-based MAC protocol for reliable broadcast in VANETs," *IEEE Transactions on Mobile Computing*, vol. 12, no. 9, pp. 1724–1736, Sept. 2013.
19. H. A. Omar, W. Zhuang, and L. Li, "On Multihop Communications For In-Vehicle Internet Access Based On a TDMA MAC Protocol," *Proc. IEEE International Conference on Computer Communications (INFOCOM)*, Apr. 2014.

Chapter 5
Conclusions and Future Works

5.1 Conclusions

VANETs are an emerging paradigm which is currently receiving significant support
from government, academia, and industrial organizations over the globe. By employing V2V and V2R communications, VANETs are expected to realize a variety
of advanced applications for road safety, passenger infotainment, and vehicle traffic
optimization. The main objective of this brief is to present a TDMA MAC protocol,
called VeMAC, to support safety and non-safety related applications in a mutichannel
VANET scenario.

In the VeMAC protocol, the nodes access the time slots on the CCH and SCHs in
distributed ways which are designed to avoid any hidden terminal problem. On the
CCH, the VeMAC provides a reliable one-hop broadcast service, which is crucial
for high priority safety applications supported on this channel. How the periodic
and event-driven safety messages are queued and served by the VeMAC protocol is
described, and a detailed message delay analysis (including queueing and service
delay) is presented by taking into consideration the size and the arrival pattern of each
type of safety messages. MATLAB simulations in highway and city scenarios show
that, compared with the ADHOC MAC protocol, the VeMAC provides a smaller rate
of transmission collisions, which results in a significantly higher throughput on the
CCH. Additionally, the network simulator ns-2 and the microscopic vehicle traffic
simulator VISSIM are used to evaluate the performance of VeMAC in comparison
with the IEEE 802.11 p standard in a realistic city scenario. Simulation results show
that, the VeMAC protocol can deliver both types of safety messages to all the nodes
in the one-hop neighbouhoud with an acceptable average delivery delay (less than
100 ms). Moreover, it is shown that the VeMAC has a low probability of a transmission collision, which results in a higher safety message goodput and better channel
utilization, as compared to the IEEE 802.11 p standard. This research sheds light
on TDMA as a promising technology for MAC in VANETs, and a suitable replacement of the IEEE 802.11 p standard, which have significant limitations in supporting
VANET safety applications.

© The Author(s) 2014   59
H. A. Omar, W. Zhuang, *Time Division Multiple Access For Vehicular Communications*,
SpringerBriefs in Computer Science, DOI 10.1007/978-3-319-09504-2_5

5.2   Further Research Topics

In the future, the performance of the VeMAC protocol using different values of the split up parameter τ (other than $\tau = 0$ and $\tau = \infty$) and the development of a scheme to adapt the τ value by taking into consideration the density of the vehicles moving in opposite directions are worth to be investigated. Another issue which must be examined is how the protocol performance is affected by the existence of asymmetric wireless channels among the nodes, as well as by the packet errors caused by the channel impairments such as noise, fading, and shadowing. Since each node interprets any packet error as a transmission collision, the packet errors due to a poor wireless channel may result in nodes unnecessarily releasing their time slots on the CCH. Also, since information security is not considered in this brief, suitable authentication and integrity schemes should be developed to protect the VeMAC protocol against any malicious attack, such as broadcasting of false control information over the CCH, which can affect the VeMAC techniques for distributed time slot assignment and transmission collision detection. Concerning the communications over the SCHs, the proposed scheme for unicast should be evaluated via analysis and simulations, and then extended to support a reliable broadcast service on the SCHs. Finally, a prototype should be created for the VeMAC protocol in order to investigate its implementation complexity and practically test its performance in a real vehicular scenario.